CONSTRUCTED WETLANDS IN THE SUSTAINABLE LANDSCAPE

CONSTRUCTED WETLANDS IN THE SUSTAINABLE LANDSCAPE

Craig S. Campbell
Michael Ogden

John Wiley & Sons, Inc.

NEW YORK / CHICHESTER / WEINHEIM / BRISBANE / SINGAPORE / TORONTO

This book is printed on acid-free paper. ♾

Published simultaneously in Canada.

ISBN 0-471-10720-4

Printed in the United States of America.

10 9 8 7 6 5 4 3 2

Preface

Wetlands—whether natural or constructed—provide a fascinating setting, teeming with myriad forms of biological life and activity. They have increasingly become recognized for their unique ecological functions in our environment and are the focus of increased research by scientists and interpretive programs by schools, communities, and nature centers.

There has been a corresponding increase in constructed wetlands and their potential to provide an effective, low-cost, natural method of removing pollutants from both wastewater and stormwater. Interest in wetlands extends from students in landscape architecture and environmental engineering programs to the real world of public officials, developers, and private citizens. Unfortunately, although there have been numerous seminars and conferences on the subject, with technical reports, proceedings, and books published almost on an annual basis, most of these publications have been in the technical realms of biology, engineering, and wetland chemistry. While this context provides a very useful and solid foundation in advancing knowledge of constructed wetlands, there has been a noticeable lack of attention to some of the most intriguing aspects of constructed wetlands—their aesthetic, landscape, wildlife habitat, and other multiple-use values. This volume is intended to provide an initial basis for understanding the unique potential that constructed wetlands can provide to integrate water renovation processes with other functions that are public amenities rather than eyesores.

The past few decades have witnessed many environmental shifts in both public values and government policies. Unfortunately, some of the most well-intended policies and regulations directed at cleanup of polluted waters have coincided with a virtual dissolution of programs that previously provided grants for new or upgraded wastewater treatment facilities to assist communities in meeting these regulations. The U.S. EPA's Construction Grant Program, which was phased out in the 1980s in favor of state revolving loan programs, provided money to promote innovative and alternative wastewater treatment systems. A variety of constructed wetlands were funded under this program, and when it was phased out, the momentum was somewhat lost. The Clean Water State Revolving Fund (CWSRF) is a loan program run by states with seed money from the EPA and other sources. This program has been the major source of funding for wastewater treatment projects, including individual homeowner loans. There are additional funding sources for on-site treatment systems that vary from state to state, and some of this funding supports pilot projects utilizing constructed wetlands in several parts of the country.

Another source of support, at a smaller scale, for grants and assistance to help communities identify, install, and monitor alternative wastewater treatment systems is the National Onsite Demonstration Project (NODP) begun in 1993 and administered by the National Small Flows Clearinghouse (NSFC), an EPA-funded unit at West Virginia University. A constructed wetland system, among others, is being tested to treat restroom wastewater at a county park site in West Virginia

The U.S. EPA established the Small Flows Research Program in 1973 to help support viable alternative wastewater treatment systems in areas where site conditions precluded the use of conventional systems. The NSFC was created under the Clean Water Act to assist small communities with their wastewater-related needs by collecting and disseminating information. *Small flow* refers to a facility with a design capacity of 1 million gallons per day or less, or to wastewater typically generated by 10,000 or fewer people. The NSFC publishes an excellent newspaper-format periodical; offers more than 600 different manuals, pamphlets, and videotapes—many addressing constructed wetlands; and offers technical assistance and outreach by means of workshops, seminars, and referrals. Their toll-free number is 1-800-624-8301, and their Web site is at http://www.nsfc.wvu.edu.

In addition, of course, at various times grants are available at state and regional levels, such as the assistance for *colonias* water and wastewater infrastructure offered by the Texas Water Development Board in border counties; Farmers Home Administration grants; and other sources of limited assistance. However, in the absence of grants, which are extremely limited in both size and availability, communities faced with increasingly stringent wastewater and stormwater treatment requirements are typically able to receive only loans—not grants—for the required improvements. This situation, which typically results in the necessity to raise sewer rates, will undoubtedly become worse for local communities, mandating consideration of the most cost-effective means of treating wastewater and other polluted water. Not only the original capital costs but also the long-term costs of operation and maintenance must be given even more attention than before.

This situation has given rise to widespread interest worldwide in utilizing 'natural systems' for wastewater treatment. Although all wastewater treatment systems, even those within the framework of 'high-tech' advanced treatment systems, utilize natural biological processes to some degree, the development over the past several decades of a knowledge base regarding wetlands processes and the ability of constructed wetland systems to renovate polluted waters has become quite sophisticated. Constructed wetlands exist in Great Britain, Canada, the United States, Mexico, India, South Africa, Brazil, Australia, and many European countries ranging in size from small single-family residential systems to major facilities treating up to 12 million gallons per day of wastewater. These systems are becoming more acceptable to local and regional regulators as the data on their performance continue to prove the cost-effective nature of their operation.

One of the most interesting aspects of the field of constructed wetlands is the multidisciplinary nature of the subject. From the very beginning, research and development within this field has attracted an admirable collection of experts ranging from chemical engineers, aquatic biologists, ecologists, and wildlife specialists to landscape architects, civil engineers, and others. The very nature of wetland processes is fascinating, and as one person put it, "addictive" the more one delves into it! At this point, the subject is in need of integration with the professional disciplines most responsible for the process of planning our communities and designing the constructed environment. To that end, we hope to engage the interests of architects, landscape architects, planners, developers, and public officials who have a clear responsibility to consider the long-range environmental costs and benefits of their efforts and decisions. This volume will therefore focus on the multiple benefits and uses of constructed wetlands and the means by which they can be successfully integrated with other community amenities such as park and open space systems, schools, museums and nature centers, wildlife refuges, and other elements of the built environment.

Aquatic plants are not unique in their capacity to provide benefits to humans and their constructed environment. Human history is so closely connected with plants and the development of their beneficial uses, ranging from food and fiber to medicines and shelter, that the Western dissociation from folk knowledge of plants over the past century appears to be an unfortunate anomaly of a culture more interested in comfort and consumption than in a deeper connection with natural processes. This aspect of Western society's recent history is undergoing a profound reorientation that is being manifested worldwide. Few countries or societies anywhere now welcome industrial development, resorts, or other high-impact projects with open arms, and there is increased interest in low-tech solutions to many of the development and pollution problems around the globe. *Sustainability* has become a key word and a critical objective for new development, and is written into the guidelines of the World Bank and most other international lending institutions for assessing new projects.

Today it is increasingly problematic to isolate or subdivide all of the various elements associated with new development and with improvements to our existing environment. Even though this volume has taken on the task of presenting one technology only—that of utilizing low-energy natural aquatic systems to renovate polluted waters—we have tried to emphasize the total interdependency of the various disciplines dealing with the built environment. At a time when most academic programs are becoming increasingly specialized—or, as the pundits have put it, "learning more and more about less and less"—the most pressing need in today's world is for more meaningful integration of diverse disciplines and areas of knowledge.

Landscape architecture is possibly the only profession that has traditionally embraced this integrative function. This stance has given the profession with a solid foundation to provide leadership on large multidisciplinary projects involving a sensitive inventory and assessment of both cultural and natural resources. Central to this role is the willingness to engage the full range of appropriate specialists and the ability to coordinate their work with minimal preconceptions.

The development of a sound corpus of theory on the development of sustainable landscapes, based on the work of Ian McHarg, Michael Hough, John Lyle, Robert

Thayer, and others, has placed landscape architecture in the lead within the field of environmental design. Unfortunately, architects are generally obsessed with theory and trendy styles, typically ignoring the myriad environmental factors that comprise a particular setting and deserve an appropriate and sensitive response. Planners, with few exceptions, have virtually abandoned the field of site planning and design, and have certainly not been responsible for the development of a body of theory encompassing guiding principles for their profession. What is now called *new urbanism* is not really new; it has been defined primarily by architects, with little reference to environmental factors.

Engineers have been strongly criticized for creating visual "monsters" that in many cases are more a result of politics, funding, and regulations than of incompetent or insensitive design. But it can fairly be said that the engineering profession as a whole has not been in the forefront in developing innovative approaches to handling stormwater runoff, parking area and roadway design, or other elements of site development within their domain. Most of the more creative aspects of stormwater detention, biofiltration, and so on have arisen from the involvement of landscape architects in the development process, often in conjunction with progressive developers of housing and commercial projects who are highly concerned with environmental issues and strongly commitmented to the principle of sustainability.

An interesting example of the emergence of a new ethic within the engineering profession and government agencies is provided by the U.S. Army Corps of Engineers. This agency, more than any other within the federal bureaucracy, was responsible for projects with devastating impacts on the natural environment. In an interesting change of direction, this agency, which has essentially run out of work in terms of new dam construction, has taken on the work of restoring wetlands habitats and re-creating the former path of the Kissimmee River in Florida, which involves backfilling the canal the Corps channeled over thirty years ago. The Kissimmee was channelized in the 1960s by the Corps to control flooding and provide continuous navigation. A 103-mile meandering river was straightened into a 56-mile canal that drained the surrounding wetlands and heavily impacted wildlife and other natural resources both in the immediate area and in the Everglades. The restoration plan, developed by the South Florida Water Management District and the Corps, is estimated to take fifteen years and cost $350 million.

There are other positive moves that recognize the importance of developing a more integrated biological approach to management of pollution problems at many levels. The objective of this volume is to further the understanding of the alternatives that deserve proper evaluation within the framework of sustainable landscapes.

ACKNOWLEDGMENTS

Thanks to the following companies, agencies, and individuals who kindly supplied information to the authors for use in preparing this book:

Aquatic and Wetland Company, Boulder, Colorado

David Brown, Environmental Engineer, U.S. EPA Risk Reduction Lab, Cincinnati, Ohio
Joanne Jackson, Post Schuh Buckly & Jernigan, Winter Park, Florida
Ron W. Crites, P.E., Nolte and Associates, Inc., Sacramento, California
Tim Darilek, San Antonio Water System, San Antonio, Texas
Edith Felchle, Division of Natural Resources, City of Fort Collins, Colorado
Matt Finn, Biosphere 2, Oracle, Arizona
Pliny Fisk III, Center for Maximum Potential Building Systems, Inc. Austin, Texas
Reese Fullerton, Tesuque, New Mexico
Carol Franklin, Andropogon Associates, Philadelphia, Pennsylvania
Peggy Gaynor, Gaynor Landscape Architects, Seattle, Washington
Michelle Girts, CH2MHill, Portland, Oregon
John Grove, Grove Constructed Wetlands, Buena Vista, Colorado
Becca Hansen, The Portico Group, Seattle
Terry Hennkens, Sewer Utility Manager, City of Columbia, Missouri
Michael Hough, Hough Woodland Naylor Dance Leinster, Toronto, Ontario, Canada
Lynn Hull, artist, Fort Collins, Colorado
James D. Hunt, P.E., Dyer, Riddle, Mills, & Precourt, Inc., Orlando, Florida
Joanne Jackson, Post Buckley Shuh & Jernigan, Winter Park, Florida
Patricia Johanson, artist, Buskirk, New York
Jones & Jones, Seattle, Washington
Lorna Jordan, artist, Seattle, Washington
Ned Kahn, artist, San Francisco, California
Robert Knight, CH2MHill, Gainesville, Florida
Allison Kukla, Phoenix Arts Commission
Camilla Rode Laughlin, Land Use Department, Boulder County, Colorado
Deborah Levy, Parks Naturalist, City of Orlando Recreation Bureau
Eric H. Livingston, Environmental Administrator, Florida Dept. of Environmental Protection
Prof. Paul Lusk, Department of Architecture and Planning, University of New Mexico, Albuquerque
Michael Maglich, artist, Phoenix
Thomas McDonald, Water Reclamation, City of Beaumont, Texas
Simon Miles, Toronto, Ontario, Canada
Randy Neill, Arkansas Department of Health
Rodney Pond, Metro Transit, Seattle, Washington
Resource Conservation Technology, Inc.
Thomas Schueler, The Center for Watershed Protection, Washington, D.C.
Michael Singer, artist, Vermont
Gerald Steiner, professional engineer, Chattanooga, Tennessee

Robert Stout, artist, Albuquerque, New Mexico

Robert Thayer, Chairman, Dept. of Landscape Architecture, University of California/Davis

Unified Sewerage Agency, City of Hillsboro, Oregon

William Wenk, Wenk & Associates, Denver

Dr. Kevin White, University of Southern Alabama

Contents

1

The Concept of Sustainable Development

There is nothing more difficult to take in hand, more perilous to conduct, or more uncertain in its success, than to take the lead in the introduction of a new order of things.

—Niccolo Machiavelli (*The Prince,* 1532)

CONSTRUCTED WETLANDS AS SUSTAINABLE DESIGN

For at least fifty years, conventional wisdom has mandated the development of extensive wastewater collection systems directed to a centralized treatment plant. More attention is now being given to the benefits of a decentralized approach to treating wastes at their source—*point source* treatment, as it is called when related to industrial wastes normally dumped into storm drains. Such an approach also has value when applied to sewage wastes and is more in line with the philosophy of *sustainable development.* In many situations, a decentralized system of treating sewage wastes, potentially with constructed wetlands, can provide not only a more economical and energy efficient means of achieving treatment objectives, but also a resource in the form of reclaimed water available for landscape irrigation or creation of wildlife habitats. Such an approach may have value both in new developments remote from existing wastewater treatment facilities and in areas in need of upgrading or retrofitting septic tank and leach field systems that are polluting the groundwater.

Conventional means of treating wastewater typically involve rather unattractive industrial-looking facilities surrounded with chain link fencing. They usually have no functions except those related to treating wastewater and do not add to the visual quality or value of the surrounding area. In fact, new conventional facilities are extremely difficult to locate and generally have a negative image. There is now at least belated recognition in some parts of the country of the importance of both architectural quality and artists in the process of designing wastewater and other infrastructure projects. While the focus of this book is on decentralized, or on-site, treatment of wastewater and stormwater, we recognize that there is little choice in many communities that have already committed to centralized systems for many years. The best option for these communities and cities is to consider carefully the visual impact of any new expansion projects, and plan for public interface and education to be programmed into any new projects from the beginning. One notable recent project is the Oceanside Water Pollution Control Project in San Francisco, designed by the firms CH_2MHill, engineers, with the architectural design leadership of Simon Martin-Vegue Winkelstein Moris (SMWM). Described as a "conceptually innovative and environmentally sensitive wastewater treatment facility," the 45-acre project site is located on the Pacific coast and surrounded by public lands such as the San Francisco Zoo and the Golden Gate National Recreation Area. Concerns over the potential impact of the project led to a design with at least two-thirds of the building area underground, in an artificial canyon, with access through a series of tunnels. The facility met the objectives related to visual impact while establishing a milestone by carefully providing a setting: a walled, terraced, semipublic garden attractive to visitors who will tour the plant. Cathy Simon of SMWM feels that the architecture and site design of this project will be part of the education process for the public and states that "at the end of the twentieth century, an ecologically responsible bulding project can also be a resonant public place, capable of interpreting an industrial process of great importance to people's lives."

There have been other projects that have gone even further to integrate the interpretive process, involving artists and landscape architects in central roles in collaborative teams. These projects are described in detail in Chapter 9.

Sustainable design has become a worthwhile objective of our society and has in fact been adopted as a goal of many professional design associations. Within academia, there is an increasing awareness of the principles of sustainability, and units such as the Center for Sustainable Communities in the College of Architecture and Urban Design at the University of Washington have been established to foster research, coordination between programs, and technology transfer.

John Lyle and others at California State Polytechnic University in Pomona have developed the Center for Regenerative Studies on 16 acres at the campus to provide a setting for research, education, and demonstration of *regenerative technologies.* With twenty residents in the first phase, the community is planned to increase to ninety, providing opportunities to grow food, recycle waste, generate energy, and, in general, "learn by doing."

While the term may have different meanings for different people, one appropriate definition, as adopted by the National Park Service, is: ". . . Sustainability as related to park planning, design, and development means meeting present needs without compromising the ability of future generations to meet their own needs. Sustainability minimizes the short and long term environmental impacts of development activity through resource conservation, recycling, waste minimization, and the utilization of energy efficient and ecologically responsible materials and procedures for construction."

Organizations representing design professionals have also made serious efforts to assist their members in assessing the environmental impacts and cost/benefits related to total life cycle costs of various materials. The American Institute of Architects (AIA) is to be commended for their intensive effort to provide a basis for selecting products and materials that have the least impact on the environment;

the organization now publishes the *AIA Environmental Resource Guide,* with quarterly installments offering detailed information on material life cycle assessments, case studies, and other information.

Wetlands have become a topic of increasing interest around the country, both to governmental regulators and to scientists, and the nature of wetlands processes has only recently begun to be understood. Constructed wetlands are increasingly being recognized as a relatively low-cost, energy-efficient, natural means of treating sewage, agricultural and industrial wastes, and stormwater runoff while at the same time offering the potential for multiple benefits. Among these benefits are the potential for integration of constructed wetlands into park and recreational systems, the wildlife habitat they provide, their aesthetic qualities, and the superior quality effluent that they produce, which can be recycled for landscape irrigation or impounded in an attractive and educational pond of value in attracting wildlife while also conveying information on wetlands processes.

In many situations, constructed wetlands can be designed to rely almost entirely on natural processes and gravity flow, thus conserving energy by minimizing or eliminating the use of pumps and mechanical equipment. In addition, the maintenance required is considerably less than with a conventional system, and the entire treatment "train" often has the ability to operate for long periods with no human intervention whatsoever. These characteristics generally place constructed wetlands in the sustainable landscape category, particularly due to their ability to provide multiple functions and benefits at low cost and with low environmental impact.

The following is excerpted from USEPA, 1988, p. 1:

> The trend over the past 70 years in the construction of water pollution control facilities for metropolitan areas has been toward "concrete and steel" alternatives. With the advent of higher energy prices and higher labor costs, these systems have become significant cost items for the communities that operate them. For small communities in particular, this cost represents a higher percentage of the budget than historically allocated to water pollution control. Processes that use relatively more land and are lower in energy use and labor costs are therefore becoming attractive alternatives for these communities. . . . The interest in aquatic wastewater treatment systems can be attributed to three basic factors:
>
> 1) Recognition of the natural treatment functions of aquatic plant systems and wetlands, particularly as nutrient sinks and buffering zones.
> 2) In the case of wetlands, emerging or renewed application of aesthetic, wildlife, and other incidental environmental benefits associated with the preservation and enhancement of wetlands.
> 3) Rapidly escalating costs of construction and operation associated with conventional treatment facilities.
>
> . . . Where natural wetlands are located conveniently to municipalities, the major cost of implementing a discharge system is for pumping treatment plant effluent to the site. Once there, further wastewater treatment occurs by the application of natural processes. In some cases, the wetland alternative can be the least cost advanced wastewater treatment and disposal alternative.

At the time when this Environmental Protection Agency (EPA) report was prepared, examples of wastewater conveyed into existing natural wetlands were more common than constructed wetlands, and the positive attitude reflected in the report was an indication of the substantial water quality improvement that had been monitored for many years within these natural wetlands. With increasing emphasis on protection of natural wetlands, this practice came to a logical halt as more information was developed on the design of constructed wetlands, which offered much more control and better opportunities for monitoring. In a number of later reports and studies, the EPA noted these same advantages in comparative life-cycle studies of constructed wetlands and advanced wastewater treatment methods demonstrating that wetlands were more economical given the availability of land at a reasonable cost.

LANDSCAPE ECOLOGY

As a basis for a theory of sustainable design on a regional or site-specific basis, landscape ecology would ideally provide designers with a body of knowledge capable of being employed, in some capacity, as a general framework. However, although landscape ecology has grown in stature through research and publications, it remains a cumbersome assemblage of theory and data from myriad specialized disciplines from the natural sciences combined with an overlay of cultural characteristics. In an ideal world, landscape ecology would help define and support the role of constructed wetlands in any setting, and would assist in efforts to mesh seamlessly with adjacent land uses and to integrate with greenways, wildlife corridors, and buffer zones.

Budgetary realities do not normally allow in-depth ecological inventories of a wider area to be undertaken as part of a constructed wetland project. Typically, however, abundant information is already available on the soil characteristics, slopes, climate, vegetation, and other aspects of a particular location to provide a reasonable basis for developing an understanding of localized landscape ecology.

One of the most useful and concise guides for designers is *Landscape Ecology Principles in Landscape Architecture and Land-Use Planning* (Dramstad et al., 1996), a slim volume with good diagrammatic illustrations of the principles underlying the concepts of edge habitat, patches, fragmentation, corridors, connectivity, and mosaics. Given the diversity of conditions affecting any localized site, especially those that are artificial, it is impossible to develop a "cookbook" equally applicable to all locations. Designers need to be sensitive to the long-range impacts of their projects, and greater emphasis on landscape ecology can provide a fuller understanding of both potential problems and opportunities for expanding the integrative functions of a constructed wetlands project.

THE SUSTAINABLE LANDSCAPE MOVEMENT

The father of the sustainable landscape, in terms of being the earliest to espouse the principles now underlying the philosophy of low-impact, environmentally sound site development, is Ian McHarg. His book *Design with Nature* has had a profound impact on the way we look at the land inventory and development process and has provided the intellectual basis for many of the more recent computerized geographic information systems (GIS) databases that are now widely employed as the basis for urban and regional land use and planning decisions. Even as these systems become more widely employed, however, one cannot ignore a finer level of specific detail, the character or genius loci of any particular site, which is impossible to digitize and classify into neat categories at the cell level of any GIS system. There is a very real danger of the technique or the technology becoming an end in itself rather than a useful tool that is carefully and judiciously employed primarily for developing broad classifications to assist in planning efforts. Potentially, such a system can be used as a convenient but superficial means of justifying improper development by an inaccurate representation of computerized models as the most "scientific" and unassailable source of information. Most of us who have had extensive experience with mapping of all types, and at all levels of detail, are surprised every time we set foot on a site on which we assumed we had all the relevant information; there are many special elements in any landscape, ranging from a single tree of great character to unique rock outcroppings or geologic anomalies, that are impossible to assess with aerial photography and GIS systems. That having been said, however, GIS systems do provide a remarkably efficient means of storing and retrieving information on ownership, vegetation, wildlife habitat, geology, slopes, zoning, and a host of other attributes that can provide at least a gross understanding of the characteristics of large areas. The weakness is evident when one needs more detailed information on specific smaller areas, including types of information neither readily measurable nor easily quantified for storage in any computerized mapping system.

The seminal project incorporating McHarg's concepts of careful and thorough analysis of environmental conditions as the basis for developing a sound plan was Woodlands, a planned community north of Houston, Texas (Figure 1.1). The level of care exercised by the developers of this community has not been adequately conveyed in most accounts of the project and has rarely been equaled since. The project was started in 1971, and the site development principles developed by the firm Wallace, McHarg, Roberts, and Todd as environmental consultants addressed the issues of varying soil permeability related to carrying capacity and potential for stormwater infiltration; surface management of stormwater; and creating a suitability model for developing the most appropriate portions of the site while retaining open space and stormwater management areas as dictated by the site and soils inventory.

Another great pioneer in the field of sustainable landscapes is Joachim "Toby" Tourbier. While a professor in the Department of Landscape Architecture at the University of Pennsylvania, Tourbier became interested in biological means of controlling water pollution in the early 1970s. In 1976, he became aware of the work of Kathe Seidel at the Max Planck Institute in Germany, along with the work of others, and organized the first major international conference on the subject of biological, low-tech methods of stormwater and wastewater renovation. In a more recent effort, *Lakes and Ponds*, a publication of the Urban Lands Institute, Tourbier and Richard Westmacott of the University of Georgia present a number of projects involving constructed stormwater wetland ponds and the biological processes involved (Tourbier, 1992).

Since then, other authors have presented a more fully developed philosophy of human impact on the landscape and on urban systems in books such as *City Form and Natural Processes* by Michael Hough of Toronto; *Regenerative Design for*

Figure 1.1 *Woodlands, Texas: Natural infiltration area (Wallace Roberts & Todd).*

Sustainable Development by Prof. John Lyle of California State Polytechnic University, Pomona; and *Gray World, Green Heart* by Prof. Robert Thayer of the University of California at Davis.

Michael Hough's book, published in 1984, presented the first integrated philosophy of bringing environmental values into the urban landscape and presented a strong, well-developed case for biological diversity within urban landscapes that has not been given its proper recognition. In a later volume directed at restoring a sense of regional identity to the landscape (Hough, 1990), the author stated succinctly the connection between sustainability and an understanding of natural systems:

> Sustainability involves, among other things, the notion that human activity and technological systems can contribute to the health of the environments and natural systems from which they draw benefit. This involves a fundamental acceptance of investment in the productivity and diversity of natural systems. (p. 193)

Pointing out that the development of a new environmental ethic of sustainability requires a major shift in attitudes, Hough points out that

> the principle of investment in nature, where change and technological development are seen as positive forces to sustain and enhance the environment, must be the basis for an environmental design philosophy. Its principles of energy and nutrient flows, common to all ecosystems when applied to the design of the human environment, provide the only ethical and pragmatic alternative to the future health of the emerging regional landscape. (p. 194)

It is interesting to note that these books, by landscape architects, have established a genuinely new theory of sustainable landscape development in a profound departure from all past design theory, which was primarily based upon aesthetics. By laying out this new framework, an ecological basis for design, they—along with others—have clearly established a foundation for practice of landscape architecture that has no counterpart in the field of architecture or city planning. In fact, both of these allied professions need to better integrate into their own worldviews the principles of sustainable landscape development to bolster their own less developed—and more static—philosophy of land planning. As John Lyle stated, "In the perspective of history, we might even view this predictive approach as a new phase in the shaping of the physical environment. Looking back, we can see at least three earlier phases that brought us to where we are now." Lyle goes on to describe, first, the *instinctive* phase, which involved building without forethought but with instinctive insight—witness the complexity of a spider web or beehive. The second phase involved the development of tradition and the application of accumulated knowledge that often achieved an admirable environmental "fit." The third phase was the *form-making* phase, typified by the application of reason, invention, and creativity illustrated on paper prior to construction. The emerging phase that Lyle describes as the one we are beginning to create and participate in involves *predictive adaptation,* which utilizes the skills developed in earlier phases and, in addition, develops our abilities to store information and predict and plan for adaptive change. Lyle warns us not to become too smug about our perceived technological abilities, quite properly recognizing that the unexpected can always happen.

BIOSPHERE 2

The earth represents Biosphere 1; Biosphere 2, by contrast, was originally developed as an experimental 3-acre glass-domed complex in the Sonoran Desert of Arizona designed to emulate the earth itself (Figure 1.2). The Biosphere 2 project achieved considerable publicity—both positive and negative—related to its experiments placing humans in a closed environment comprising miniature oceans, marshes, rain forests, and 3,800 species of animals and plants. This project was advertised as the largest self-sustaining ecosystem in the world, but in reality it represented from its inception an example of bad science combined with almost unlimited financial resources. While the concept of constructing a wide range of miniature ecosystems within a huge greenhouse in Arizona had considerable potential from the very beginning as a research and educational facility, the concept of the experiment as representing any type of model for space colonies (the underlying principle guiding the entire project) was always farfetched.

Figure 1.2 *Biosphere 2—view of the facility (Decisions Investments, Inc.).*

The designers of the facility recognized the need to process the system's wastewater naturally and to recycle both the water and the nutrients for use in a closed system. Drawing from the bioremediation experiments of Billy Wolverton at the NASA Stennis Space Center in Mississippi (Wolverton, 1989), the facility's designs included a wastewater treatment and recycling system capable of handling the sewage of up to ten people and their domestic animals by utilizing wetlands in a series of tanks. The first tank in the series contains a gravel substrate with an overlay of soil planted with cattail, bulrush, reeds, and tall grasses. This unit is essentially a subsurface flow unit, which then feeds into a second tank also planted with aquatic plants but with exposed water in open, meandering channels. The third tank is more of a pond and has small islands; from there, the water is pumped into a utility water tank, where it is stored for use in irrigating agricultural crops. All three tanks together take up only about 300 square feet and process all of the wastewater from the toilets, showers, and sinks produced by the humans occupying the structure (Figure 1.3).

The effluent, following anaerobic digestion in a series of three 1,900-liter septic tanks, follows a meandering path through the three lagoons by a series of baffles. The perforated pipes supplying effluent from the septic tank to the primary lagoon had to be switched routinely to a second line to allow clearing of sludge and plant roots. It is worth noting that among the many luxuries that the eight original biospherians had to forgo was toilet paper to avoid overloading the wastewater treatment system. A separate marsh-type unit was installed to treat laboratory wastewater and all of the drainage from the Biosphere 2 machine shop. Potable water is produced through condensation in the ten air handler units in the habitat's basement. This system is capable of producing more than 9,500 liters per day and is either disinfected with an ultraviolet unit for use in human consumption, sanitation, and food preparation or mixed with the agricultural irrigation system water for watering rice paddies, vegetable plots, worm beds, fish culture tanks, and house

Figure 1.3 *Biosphere 2—treatment tanks (Decisions Investments, Inc.).*

plants. The two final water products from the recycling system are high-purity condensate for human and animal consumption and utility water of greatly varying quality for crop irrigation, animal pen washdown, and toilets. The eight toilets are flushed with the utility water and equipped with condensate water–supplied bidets that allow the elimination of toilet paper. There are also twelve sinks, including the kitchen and laundry, seven showers, and two washing machines, all supplied with condensate water. The average daily production of wastewater when occupied is 1,500 liters per day.

A number of problems beset the Biosphere 2 experiment during its first several years, including a buildup of gases such as nitrous oxide and carbon dioxide, excess salts in the recycled water system, and other conditions that forced the new managers to "flush out" the entire system in 1994, providing a "reset" to allow the initiation of a revamped scientific focus with a terrestrial emphasis and fewer theatrics of questionable scientific value. Some of the first experiments under the new management, which includes a consortium of scientists from Harvard and Stanford universities, will focus on a study of photosynthetic ranges of a wide range of plants with varying carbon dioxide levels. Interestingly, it was determined that excess organic material in the structure's soil set off an explosive growth of oxygen-eating bacteria, which produced carbon dioxide, much of which was absorbed by the 110,000 square feet of exposed concrete that lines the structure's interior.

THE CENTER FOR MAXIMUM POTENTIAL BUILDING SYSTEMS

Better known as "Max's Pot," the Center for Maximum Potential Building Systems was established by Pliny Fisk III in 1975 in Austin, Texas, as a nonprofit education, demonstration, and research organization to examine sustainable design and construction practices to meet the needs of individual home builders, as well as planners and developers. One of the Center's showcase projects is a demonstration farm complex in Laredo, developed in cooperation with the Texas Department of Agriculture, that highlights recycled and agriculturally based building materials. Another is the Green Demonstration Building Project in Austin and Laredo. Several of the demonstration projects have integrated on-site constructed wetland wastewater treatment systems and utilize solar energy, straw bale construction, and other low-impact technologies (Figure 1.4). One of the major applied research projects at the Center is the investigation of the potential for a wide variety of earthen materials for construction, including adobe, caliche, laterite, flyash, natural pozzolan cements, and alumina clay brick.

One of the Center's most interesting innovations is based upon the recognition that the majority of the world's population lives on or near an ocean coast and that reliable sources of fresh, potable water are in decline. Because a large percentage of potable water in most industrialized countries is flushed down the toilet, the Center designed a method for utilizing salt water, pumped from a shallow on-site well, for toilets combined with a constructed wetland treatment system for the Center for Wetland Studies, a facility in Baja California Sur, Mexico, operated by

Figure 1.4 *Wetland treatment cell, Center for Maximum Potential Building Systems (Center for Maximum Potential Building Systems).*

the School for Field Studies (Figures 1.5 and 1.6). Using a gravity flow system, standard septic tanks with effluent filters are used for primary treatment. From there, the outflow feeds into a subsurface flow constructed wetland planted with halophytes, or salt-tolerant plants, that are also hydrophytes capable of withstanding inundation. One of the plants utilized is giant reed, or carrizo *(Arundo donax).*

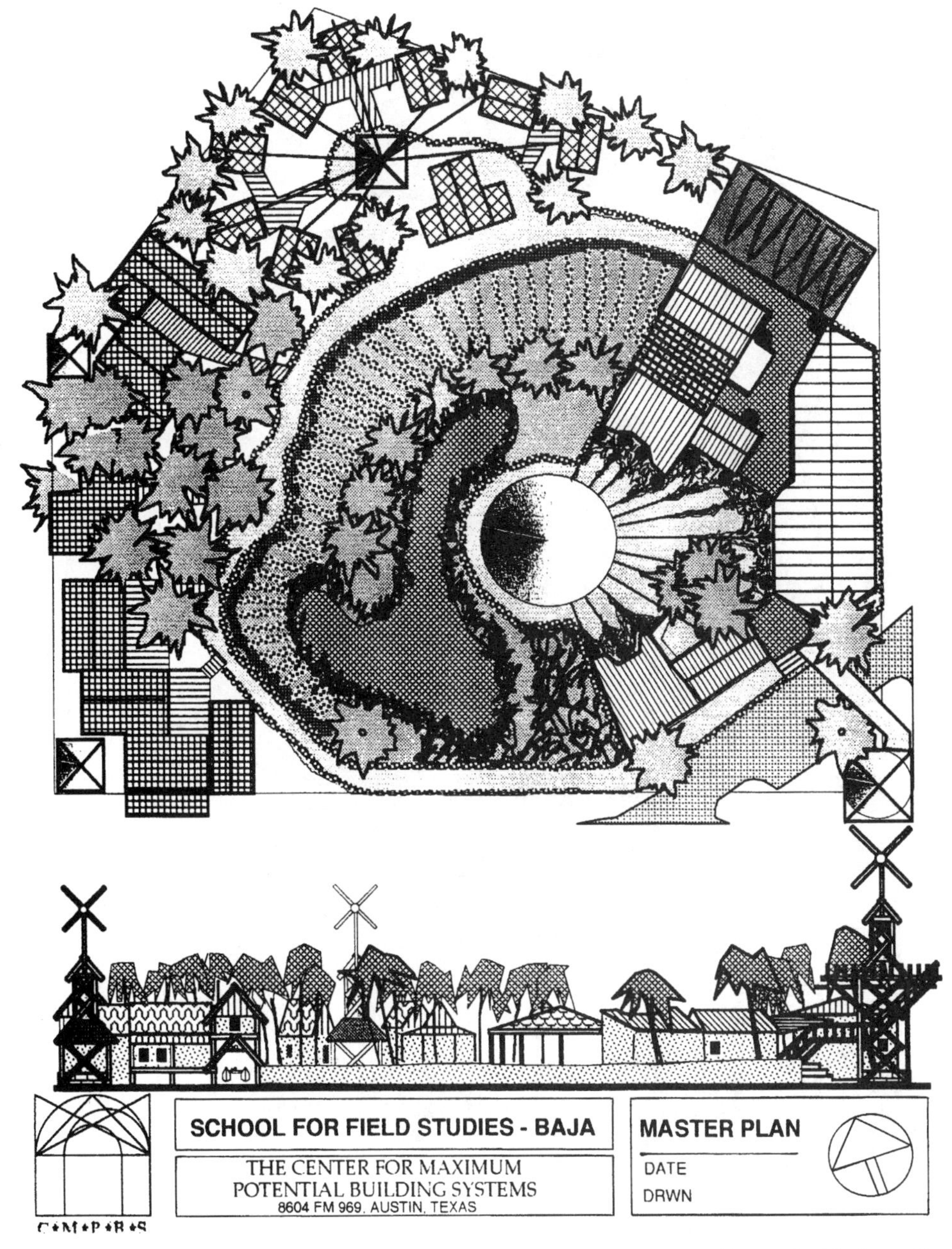

Figure 1.5 *Master plan for SFS campus with wetland treatment systems (Center for Maximum Potential Building Systems).*

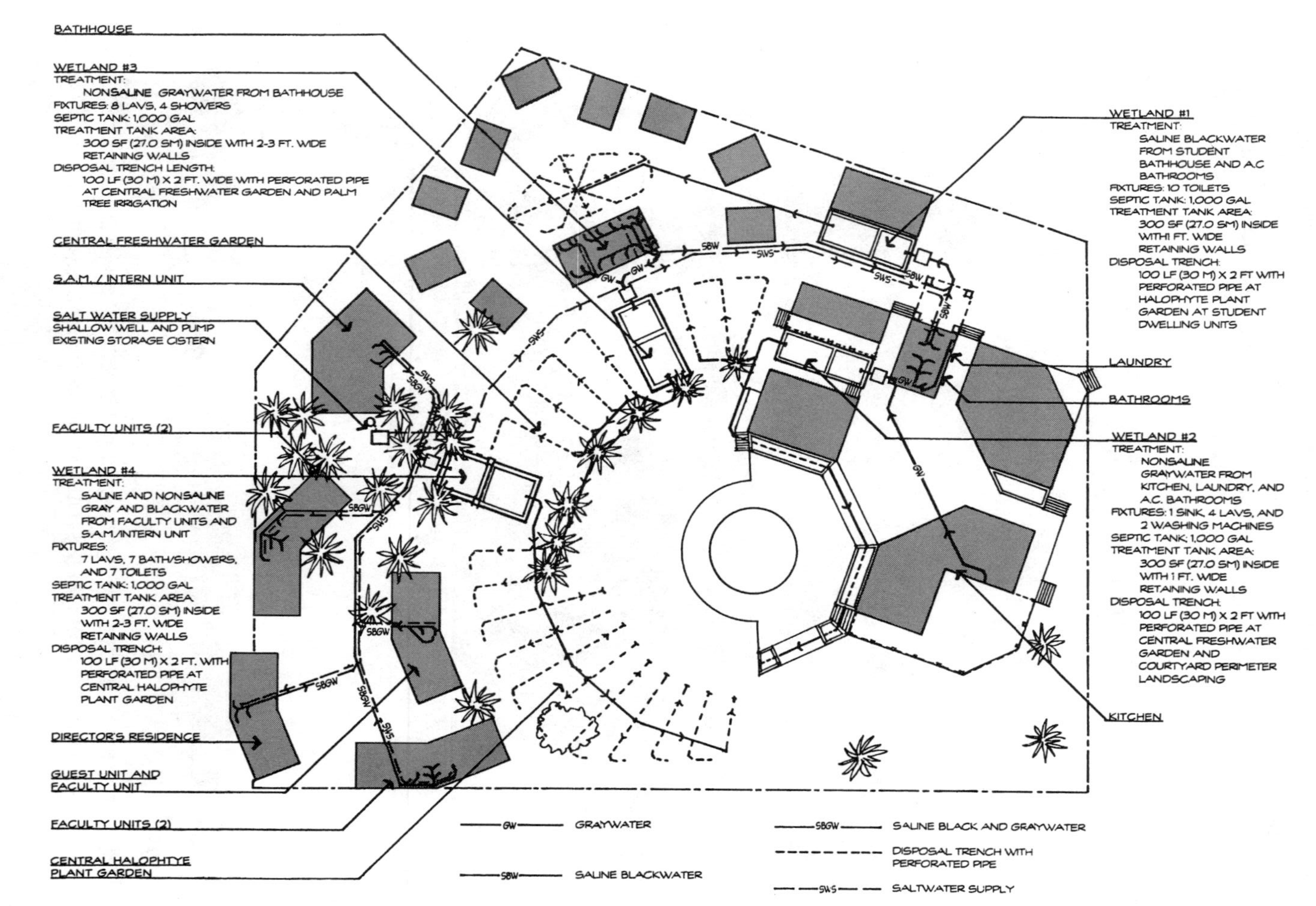

Figure 1.6 *Concept for a saltwater system, Baja California Sur, Mexico (Center for Maximum Potential Building Systems).*

There are four wetland treatment cells: two to treat nonsaline graywater from the kitchens, laundry, and bathhouse; one to treat saline wastewater from toilets; and one to treat a mixture of saline toilet wastewater and nonsaline sink, bath, and shower water. Some of the saline wastewater effluent is to be used to water ornamental halophyte plants installed to provide privacy screens between student cabins. Constructed in 1998, this project should prove to be a valuable experiment worthy of exploring in more detail in other areas as performance data are collected (Figure 1.7).

CONSTRUCTED WETLANDS AND PERMACULTURE

As stated by Mollison (1990, p. ix), "Permaculture (*perma*nent agri*culture*) is the conscious design and maintenance of agriculturally productive ecosystems which have the diversity, stability, and resilience of natural ecosystems. It is the harmonious integration of landscape and people providing their food, energy, shelter, and other material and non-material needs in a sustainable way. Without permanent agriculture there is no possibility of a stable social order." This is a high-minded,

Figure 1.7 *Baja California Sur project—student cabins (Center for Maximum Potential Building Systems).*

sensitive, and commendable philosophy to which one can hardly object; in fact, constructed wetlands fit in admirably with the overall permaculture philosophy. There is a problem, however, with the widespread proliferation of overnight permaculture "experts" whose overall knowledge of the environment is rather shallow. One of the unfortunate and dismaying characteristics of this otherwise sensible movement is the fact that it has become virtually a cult, with a high level of zealotry among its followers.

Very few individuals are capable of developing an in-depth understanding of more than one significant ecosystem or local environment, let alone the myriad range of ecosystems existing worldwide. Permaculture represents a set of concepts that are entirely laudable. What permaculture does *not* represent is a grand, innovative, new vision or philosophy of development and management of the land and its resources. None of the concepts making up permaculture are new or original; virtually all of the ideas of energy conservation—small-scale integrated food production, capturing of roof and other stormwater runoff, recycling, use of indigenous plants, and so on—have been around for centuries, with more emphasis within some cultures and at some periods of time than others.

This having been said, it is fair to assess the emergence of a high level of interest in concepts of conserving water, soil, and biodiversity as a very positive change in our global outlook. Specialists in every field, from aquatic biology to civil engineering, are more keenly aware of the interrelationships between consumption, conservation, recycling of resources, and stewardship of our land, and more serious research and interdisciplinary interaction are now taking place than ever before.

Our urban landscapes have suffered in this century in failing to achieve a connection with the natural world. The people most responsible for the design and development of urban areas have been builders and developers, engineers, politicians, planners, and architects—probably in that approximate order. Although landscape architects have had an impact on parks and streetscapes, they have typically been involved only after the basic framework was already established by others. Notably absent have been biologists, ecologists, and other natural scientists, whose perspectives are sorely needed to connect the urban pattern in a sensitive way to the natural world in which it resides. As John Lyle points out, the real challenge is to "get sustainability out of the fringe and into the mainstream where these ideas have to be if they're to have any effect on this society."

BIOREMEDIATION AND PHYTOREMEDIATION

The field of bioremediation—and, more specifically, *phytoremediation*—is one more area of endeavor closely tied to the principles of constructed wetlands for water renovation and the broader principles of sustainability. The scientific community appears to be on the threshold of a world of incredibly interesting discoveries on specific plants, bacteria, fungi, and algae, and the ability of specific species or strains to remove and break down a wide variety of contaminants.

Researchers at Ohio State University developed a method of removing cadmium from wastewater by using bacteria called *Zooglea ramigera* trapped in tiny artificial beads. With this method, this heavy metal is removed from effluent before it leaves a manufacturing plant to enter a sewage system. At the University of Hawaii, researchers report that the fungus *Penicilium digitatum* can absorb uranium from solutions of uranyl chloride. Another interesting recent event is the development by scientists at the University of California at Riverside of a method to speed the action of several types of natural fungi in converting selenium into a harmless gas. In tests at the Kesterson National Wildlife Refuge, it was demonstrated that the fungi were able to remove as much as 50 percent of the selenium in as little as four months.

There are undoubtedly many other efforts underway that may provide a wide menu of options for using specific bacteria and fungi in combination with wetland treatment systems to provide wastewater renovation—at least in smaller systems—superior in quality to that of most tertiary-level mechanical treatment facilities.

In research laboratories, academic institutions, private bioremediation companies, and other facilities, intense efforts are underway to develop specialized bacteria and fungi to clean up aquifers, toxic dumps, and oil spills. Methane-eating bacteria have been discovered and developed that produce enzymes capable of degrading more than 95 percent of contaminants such as vinyl chloride and other poisons into salt. Bioremediation techniques utilizing bacteria are cheaper than incineration and do not produce a toxic ash. Firms specializing in this work, such as Ecova, Inc., Biotrol, CAA Bioremediation Systems, and Environmental Protection North of Germany, have nurtured specialized bacteria that have been used to degrade oil, benzene, toluene, creosote, phenol, and herbicides at various sites around the earth, with high rates of success. Even resistant compounds such as polychlorinated biphenyls (PCBs) have been successfully degraded with specialized bacteria. Now there are even products on the market designed for homeowners and others who would like to remove oil, gas, transmission and brake fluids, and solvents from paving by using bacteria. The microbes are combined with a clay or vegetable extract and remain inert until activated.

Under the general heading of phytoremediation (*phyto* means "plant"), teams of researchers at Brookhaven National Laboratory, the U.S. Army Corps of Engineers Waterways Experiment Station (WES) and the National Risk Management Research Laboratory of the U.S. EPA are at several locations testing and documenting the appropriate uses of phytoremediation in constructed wetlands for its ability to degrade TNT and cyclonite contaminants in groundwater. They have determined that the best performance was achieved with elodea, sago pondweed, and water stargrass (Best et al., 1997). In the future, there may be opportunities for serious research on the development of specialized microbes that may enhance the ability of constructed wetlands to degrade some pollutants.

Toward that end, the U.S. Army Environmental Center (USAEC) has been testing the feasibility of using selected wetland plants to clean up explosives-contaminated groundwater at the Milan Army Ammunition Plant in Tennessee as an alternative to the labor-intensive, costly processes of granular activated carbon and advanced oxidation. These efforts were stimulated by tests undertaken by the EPA National Exposure Research Laboratory in Athens, Georgia, that identified a plant nitroreductase enzyme shown to degrade TNT, cyclonite (RDX), and high-melting explosive (HMX) in concert with other plant enzymes. Further testing identified similar nitroreductase activity in a wide variety of aquatic plants, which opens the door to a variety of potential applications for explosives residue cleanup utilizing constructed wetlands.

Another example of the increasing interest in the potential for plants to function as natural "remediators" of polluted soils and waters comes from the Great Plains–Rocky Mountain Hazardous Substances Research Center (HSRC), which confirmed that poplar trees (*Populus* spp.) can be utilized to prevent pesticides, herbicides, and fertilizers from contaminating surface and groundwater. As reported in EPA's *Ground Water Currents* (U.S. EPA 1993c), poplars and other plant species are also being studied at Superfund sites for use in removing other organic contaminants and metals. A three-year-old poplar crop planted along a stream bank adjacent to a corn field by a University of Iowa research team reduced nitrate nitrogen levels in leachate from fertilized fields through uptake of soluble inorganic nitrogen and ammonium-nitrogen and their conversion into protein and nitrogen gas. In addition, the trees were shown to slow the migration of volatile organic chemicals and to transform atrazine into carbon dioxide.

These are just a few examples of the shift in focus during the past half century toward belated recognition of the incredibly varied physiology and chemistry associated with plants and their potential value as sources of natural assistance in pollution control, not to mention their ever-increasing importance as potential sources of new drugs.

PUBLIC ATTITUDES AND SUSTAINABILITY

The citizens of many communities are often the visionaries, and are ahead of most government officials in their interest in developing sustainable landscapes, innovative experiments, ecological art, and collaborative multidisciplinary efforts. It is somewhat understandable that government officials prefer dealing with projects with simple mandates rather than multiple functions, and with one set of consultants rather than a collaborative effort involving various disciplines. Government standardized, single-focus projects, which often, incidentally, present roadblocks to interdisciplinary teams and experimental designs, are often mundane projects that do little to advance human knowledge or to engage people's emotions and intellect. Much of the impetus for more creative approaches to problems of infrastructure development, management of resources, and environmental preservation is coming from community groups and organizations.

Once again, it is only fair to point out that from a government administrator's or regulator's point of view, outsiders are not the ones who have to take the heat when an innovative project goes wrong. They *are*—which tends to reinforce the tendency for government agencies to support the most conservative approach to major infrastructure projects. This situation is changing, as the governmental entities that have taken the lead in sponsoring innovative solutions, from support of ecological art to constructed wetlands for stormwater renovation, have won awards and received publicity that makes the path easier for others to follow. Examples cited in later chapters of this book illustrate how some cities and other entities have taken the lead in supporting innovative multipurpose projects in a variety of venues that have created a positive image for constructed wetlands, along with a challenge to both professional designers and their clients involved in development projects around the country. A detailed description of many of these projects is presented in Chapters 9 and 10.

2

The Nature of Wetland Processes

> The marshes of the world and their related habitats—swamps, bogs and fens—are very curious halfway houses, extraordinary amalgams of land and water. . . . Most fresh wetlands harbor an extraordinary variety of life, because after all they offer the best of both worlds to plants and animals—plenty of water and plenty of sunshine.
>
> —Gerard Durrell (*A Practical Guide for the Amateur Naturalist,* 1988)

Until very recently, wetlands ecology was not a subject area that attracted much attention or support within either the academic community or federal agencies. A larger body of literature and a longer tradition of interest in wetlands exist in Europe, and much of the literature on the subject prior to the mid-1970s originated there.

One of the first major U.S. conferences on wetland ecology was held at Rutgers University in 1977 under the title "Freshwater Marshes: Present Status, Future Needs." This conference brought together forty scientists to discuss the current state of knowledge at the time, along with the management potential of freshwater wetlands. The proceedings of this conference represented one of the first collections of scientific articles focused on wetlands ecology. Since that time, of course, there has been a virtual explosion of interest in wetlands, both at the academic level and within government agencies. Annual conferences are held, and the Society of Wetland Scientists, formed in 1980, now boasts a membership of 4,000, with well-attended conferences where wetlands research is presented and discussed.

One problem facing anyone attempting to study wetland ecology and processes is the complex nature of the processes and the great variety of wetlands in different parts of the country. There are many legitimate avenues to classification of wetlands that may be derived from aquatic biota, soils, hydrology, and even watershed relationships. The high degree of interaction between functions of hydrology, soil chemistry, nutrient recycling, habitat, and so on that is characteristic of wetlands has attracted the interest of a greater variety of disciplines than perhaps any other type of ecosystem within our environment. The complex nature of the interactions occurring within a wetlands ecosystem does not lend itself readily to specialized research by one single discipline.

THE NATURE OF WETLANDS

Wetland processes are among the most complicated sets of soil and water chemistry, plant, and hydrology interactions occurring within any ecosystem on earth. The variations from one type of wetland to another are enormous, and are thus treated separately at some length in the most serious studies of wetlands, such as that of Mitsch and Gosselink (1993). The main broad categories of North American wetlands, as described by these authors, are as follows:

- Freshwater marshes
- Northern peatlands
- Southern deepwater swamps
- Riparian wetlands
- Coastal wetland ecosystems
    - Tidal salt marshes
    - Tidal freshwater marshes
    - Mangrove wetlands

The U.S. Fish and Wildlife Service began a national inventory of wetlands in 1979 that was approximately 85 percent complete in 1998. The effort involves mapping at a gross scale and identifying the basic characteristics of each wetland identified. The Fish and Wildlife Service has also developed, in cooperation with regional botanists and aquatic specialists, lists of wetland plant species occurring within each of their regions.

FUNCTIONS AND VALUES OF WETLANDS

The multiple functions and values of wetlands have only recently begun to be recognized. Previously, there were few centers of research or university programs focusing on the study of wetland ecosystems, which are among the most important ecosystems on earth and which have been disappearing at an alarming rate. Due to the incredible diversity exhibited by wetland ecosystems around the world, they have been difficult to define precisely. Hydrology, botany, aquatic biology, soil chemistry, and microbial biochemistry are areas of specialization necessary to understand fully the ecology of wetlands. When we enter the "applied" field of constructed wetland design, we must add the talents of the landscape architect and the civil engineer.

The following are functions and values of wetlands, as recognized by the Wetland Evaluation Technique currently used by the U.S. Army Corps of Engineers and other agencies (Adamus et al., 1987, 1990):

1. Functions: (which may also be considered values to some)
   a. Groundwater recharge
   b. Groundwater discharge
   c. Floodwater alteration
   d. Sediment stabilization
   e. Sediment/toxicant retention
   f. Nutrient removal/transformation
   g. Production export
   h. Aquatic diversity/abundance
   i. Wildlife diversity/abundance
2. Values: (which do not perform functions within the wetland)
   a. Recreation
   b. Uniqueness/heritage values

Not included in this list (unless indirectly implied within "Uniqueness") are values associated with aesthetics or visual beauty, which admittedly may not apply to all wetlands but which certainly are a significant part of the value of most wetlands in the West.

It has been proven that the highest productivity occurs in wetlands that have the highest flow-through of water and nutrients, or in wetlands with a "pulsing" hydroperiod. Since both of these characteristics typify constructed wetlands for wastewater treatment, it is not surprising to observe extemely vigorous aquatic plant growth after the second year of installation, which generally surpasses that of natural wetlands with less nutients or flow-through. Mitsch and Gosselink (1993, p. 55) state that "Hydrology is probably the single most important determinant of the establishment and maintenance of specific types of wetlands and wetland processes." They point out that although soil biochemistry and microbial communities are important, these and many other wetland constituents are ultimately controlled or enhanced by hydrologic conditions, and wetland ecosystems are continually responding to their hydrologic conditions. Hydrology is extremely important in determining the character of a natural wetland. The effects of the *hydroperiod,* which is the seasonal pattern provided by the rise and fall of a wetland's surface and subsurface water, determine the nature of the vegetation, along with many other factors.

Since the biotic component of a wetland, mainly vegetation, can affect water conditions through many mechanisms, including shading, sediment deposition, transpiration, and so on, there are a number of unpredictable long-range effects that may develop as a constructed wetland evolves. One factor that has not been monitored over a period of decades is solids deposition. There is evidence that to some degree, solids are decomposed and do not ultimately result in sedimentation of a constructed wetland; however, there must be a sufficient carbon source to allow adequate decomposition to take place, either in the plant vegetative litter or in the solids that may flow into the wetland from a pretreatment unit such as a septic tank or lagoon. To some degree, colloidal particles such as suspended solids are attracted to aquatic plant root hairs due to the opposite electrical charges. These particles adhere to and are slowly digested and assimilated by the plants and microorganisms.

Flood-tolerant plant species (hydrophytes) have evolved with a set of adaptations that allow them to tolerate a range of stress conditions that would result in shock or destruction of most upland plants. One of the primary adaptations that many wetland plants exhibit are air spaces (*aerenchyma*) that allow oxygen to be transported from the upper parts of the plant into the roots. These can clearly be seen if a cattail stem is cleanly severed (Figure 2.1). Studies by wetland scientists have proven that not only do many emergent wetland plant species supply oxygen to the roots, but that they also diffuse out and oxidize the adjacent soil. There appear to be considerable variations in the extent of oxygen leakage along the length of the roots, with water lilies leaking oxygen only from the root apex. The soil redox potential of these environments allows otherwise toxic ions such as manganese to be reoxidized and precipitated in the soil. Test results from various constructed wetland facilities utilizing cattail and bulrush plants indicate that cattail species may typically send roots down only about 1 foot (0.3 m) in depth, thus allowing much of the wastewater flow in a subsurface flow wetland to pass through the bed below the root zone. Bulrushes, on the other hand, have the ability to extend their roots down to the bottom of a 2.5-foot (0.76-m)-deep bed (Figure 2.2). There would appear to be a direct correlation between the ability of a subsurface flow wetland to remove ammonia and the depth of root penetration, as enough oxygen must be present within the gravel bed to facilitate denitrification. In a subsurface flow installation in Bear Creek, Alabama, utilizing cattail in a 1-foot (0.3-m)-deep bed with roots extending to the bottom of the bed, ammonia removal reached 80 percent; in a 2.5-foot ((0.76-m)-deep bulrush bed with roots to the bottom at Santee, California, ammonia removal reached 94 percent (U.S. EPA, 1993c).

Figure 2.1 *Cattail—aerenchyma tubes (Craig Campbell).*

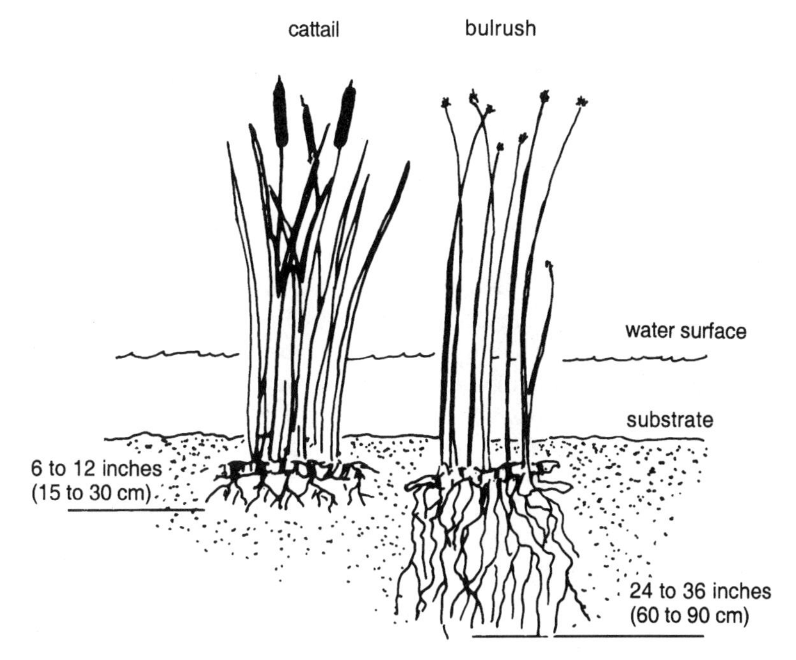

Figure 2.2 *Typical rooting depth—cattail and bulrush.*

AQUATIC PLANTS AND WASTEWATER RENOVATION

Wetlands have long been used as waste disposal sites. Until fairly recently, the unique qualities of wetlands and their net primary productivity were poorly understood. The characteristics of mosquitoes and other insects, mucky ground, and other aspects of wetlands contributed to their negative image. The purification potential of water by wetlands, however, has been recognized for centuries, and the natives of Sudanese villages along the Nile have long used indigenous wetland plants and clay soils to purify water from the river during the flood season.

Aquatic plants play a much more important role in human history than is commonly thought. Rice, of course, is the aquatic plant that has played the most important role in the health and culture of a great proportion of humanity throughout the centuries. Lotus is another important plant both symbolically and as a source of food in the Orient. In this country, of course, wild rice was historically a major foodstuff for the Ojibway Native Americans in the upper Midwest lake country. Cattail tubers were utilized as a foodstuff by many other tribes. Watercress (*Nasturtium officinale*) is extremely nutritious, and the Romans believed that it could quiet a deranged mind. Water lilies were harvested and cultivated for human consumption in ancient Egypt. The method of rolling water lily seeds in balls of clay and spreading them over the water to ensure a future crop originated in ancient Egypt and was still recommended in the early part of this century in this country. Papyrus was used as a source of pulp for paper by the Egyptians, and various reeds have been tied in bundles for making boats and are still used on lakes in Bolivia. Common reed (*Phragmites australis*), which occurs in vast areas of the world in both temperate and tropical climates, is viewed as a pest in parts of the United States, where it can quickly colonize large areas of wetlands. It has been widely used in other countries for thatching roofs, making fences, and other uses, and in Romania it is grown as a commercial crop in the Danube delta, where special machines harvest the reeds for processing in pulp mills that extract the fiber for use in paper, cardboard, cellophane, and other products. Additional uses, as reported by Riemer (1984), are insulation and fiberboard, alcohol, and even fertilizer. Common reed has now become established as one of the primary aquatic plants used in constructed wetlands in Denmark, Great Britain, and other European countries, where such installations are called *rock reed filters.*

A unique range of biological life forms have adapted to and inhabit wetlands, some of which have a special role in constructed wetlands habitat. From the unicellular protists such as *Nitrosomonas*, bacteria that are actively involved in oxidizing ammonia, to plants, birds, reptiles, fish, and mammals, the variety of life forms inhabiting the ecological tapestry of wetlands is quite amazing and of great interest to children as well as naturalists of all ages. As previously mentioned, aquatic plants such as cattails and bulrush have developed tubular pore spaces in their stems and leaves that allow the transport of oxygen to the root zone. This oxygen transport increases the redox potential of the substrate, rendering it more favorable for root growth and altering its chemical behavior.

Traditional wastewater treatment usually relies on the activity of bacterial microbes, both anaerobic and aerobic, for much of the treatment process. Each type of bacteria plays a particular role at different stages of wastewater treatment. What is lacking in conventional wastewater treatment are processes that provide an environment for both anaerobic and aerobic microbes to thrive in the same area, an ideal situation for the most efficient nitrification and denitrification processes to occur. A constructed wetland, however, does provide such an environment, with an anaerobic zone surrounding the root zone that at the same time provides a mini-aerobic zone surrounding the root hairs that fix the oxygen pumped down by the stems and/or leaves of the aquatic vegetation. The process of nitrification converts nitrogen compounds into the nitrate form; the process of denitrification transforms the nitrate into a gaseous form so that it can be eliminated into the atmosphere. Denitrification is the microbial conversion of nitrate to gaseous nitrogen, which results in the effective removal of nitrogren from wastewater.

Metals in soluble form such as iron, manganese, and copper are typically transformed by microbial oxidation and precipitated in the wetlands substrate in the form of oxides or sulfides. In summary, constructed wetlands can significantly reduce biological oxygen demand (BOD_5), suspended solids (SS), and nitrogen, as well as metals, trace organics, and pathogens. The basic treatment mechanisms include sedimentation, chemical precipitation and adsorption, and microbial interactions with BOD_5, SS, and nitrogen, as well as uptake by the vegetation.

Although nearly 5,000 plant types may occur in wetlands in North America—plants known as *hydrophytic vegetation*—most aquatic plants fall into three categories. Aquatic vegetation, which grows completely below the water surface, is classifed as *submergent* (*Elodea,* etc.); plants that are rooted in the soil but send stems and leaves above the water are *emergent* (cattail, bulrush, etc.); and those in which the entire plant floats freely on the water surface are classified as *floating* (duckweed, water hyacinth). We are primarily interested in the emergents for purposes of wastewater renovation due to their adaptability, their deep root structure, and their local presence.

One example of a floating plant is the water hyacinth (*Eichhornia crassipes;* Figure 2.3), a native of South America, which was originally introduced into the United States as an ornamental plant in a pool at a Cotton States Exposition in New Orleans in 1884 and has since gained notoriety as a prolific pest due to its explosive reproductive rate. That same characteristic, combined with the morphology of the plant, allows it to remove contaminants from water at an impressive rate. It has been widely utilized in Florida in constructed wetlands for wastewater treatment, but is considered an invasive plant and is not hardy in areas with freezing winter temperatures. It is a plant with thick leaves in a rosette form, with extremely dense roots extending down into the water that have the ability to absorb pollutants very effectively. Water hyacinth has also been utilized in a large number of experimental pilot-scale wastewater treatment projects, especially in California, and is employed extensively by the City of Austin in its Hornsby Bend sludge dewatering facility within a huge greenhouse structure to assist in removing contaminants from the sludge stream while also providing a source of bulking and carbon for their composting operations.

Duckweed (*Lemna, Spriodela, Wolffia, etc.*) is a tiny floating plant that has also been employed in wastewater renovation, primarily within a patented system of floating cells and mainly in southern states. Due to its small size—it is the smallest flowering plant—it will drift in windy conditions and thus must be contained within floating barriers. Duckweed can form a solid mat, thus reducing light penetration and algal growth in lagoons. It must be harvested annually and thus involves special equipment and maintenance considerations. The most common species of duckweed employed in wastewater treatment lagoons is *Lemna minor,* due to its extremely vigorous growth, which can make it a troublesome weed in aquatic plant nurseries if it is not wanted. The reproductive rate of duckweed is amazing; it has

Figure 2.3 *Water hyacinth (Craig Campbell).*

been estimated that if left undisturbed, a 1-square-inch (645-mm^2) patch of *Lemna minor* would grow to cover well over an acre in fifty-five days! Another species that is more suitable for ornamental ponds is *Spirodela polyrrhiza,* which does not grow as fast as most *Lemna* species. All duckweed species effectively take up nutrients from water, and are relished by waterfowl and some species of herbivorous fish.

Some submergent species of aquatic plants such as *Elodea, Hydrilla,* and *Egeria* are capable of taking in all the nutrients they need with the entire surface of the plant body. Although they may be rooted in the substrate, they are not dependent on it. They are able to thrive unattached and floating, and are still capable of growing and branching.

Many emergent aquatic plants have rhizomes that branch out and form thick clusters that can be split into smaller sections for replanting. Among the plants with such rhizomes are *Iris pseudocorus, Acorus calamus, and Peltandra* species.

The objectives of constructed wetland design are heavily influenced by their primary function, which is to remove contaminants as completely as possible. The plant selection for the treatment portion of the wetlands, therefore, is influenced by different factors than those that may guide wetland mitigation or restoration projects. Normally, a much smaller group of aquatic plants are employed in constructed wetlands for wastewater treatment than those intended for stormwater treatment or for wetlands intended solely as restoration or mitigation features that would normally require the widest diversity of plantings. Several objectives have influenced the narrower selection of plantings for constructed wetlands:

- The plants are basically emergent species capable of withstanding water depths ranging from 6 to 24 inches. Thus they can be utilized in either surface flow or subsurface flow wetlands.
- The plants should exhibit vigorous rooting that extends both laterally and vertically; the longer the vertical root depth, the better in terms of contact area for microbial bacteria and for introduction of oxygen into an otherwise anaerobic root zone.
- The plants should be capable of reproducing and infilling rapidly.
- The plants are available preferably locally

There are three primary emergent aquatic plants that have had the widest application in constructed wetlands, due in part to their vigorous growth and rooting habits and in part to their widespread availability.

Cattails (*Typha* spp.) occur worldwide in the Northern Hemisphere and are found in varying forms throughout the United States. They are vigorous growers, are capable of thriving under diverse environmental conditions, and are easy to propagate. Rhizomes (a thickened root) can be collected and planted and will produce plants in one growing season. As previously mentioned, cattail species are not likely to extend roots down to a depth greater than 1 foot and are thus not as efficient as bulrush in oxygenating a deeper gravel bed (Figure 2.4).

Figure 2.4 *Cattail (Craig Campbell).*

Bulrushes (*Scirpus* spp.) also grow in a diverse range of both inland and coastal waters, and various species are found throughout most of the United States in wetlands and in lake or pond shallows. While not as vigorous or widespread as cattails, bulrushes are very efficient in removing nitrogen and tolerate a wide pH range. Bulrush species appear to have roots capable of penetrating to a depth of 2.5 to 3 feet or greater and are thus extremely useful in oxygenating the deepest portion of a gravel subsurface flow bed (Figure 2.5).

Common reeds (*Phragmites australis*) are tall annual grasses with extensive perennial rhizomatous roots that typically penetrate to a depth of 18 inches. Their height ranges from 6 to 12 feet, with flowers of spikelets in July to October. They are attractive plants, quite lush in appearance, and provide a good background when height is needed. Reeds have been extensively utilized in Europe in the *root zone* method of wastewater treatment and are very effective in transferring oxygen due to the depth of penetration of the roots (Figure 2.6).

Due to these desirable characteristics, cattail, bulrush, and reed species have been the most favored plantings for treatment cells in constructed wetland projects. Both cattail and bulrush offer several different species that may be appropriate

Figure 2.5 *Bulrush (Craig Campbell).*

within different regions, and both tend to grow and fill in quickly. In the Southeast, in states such as Alabama and Mississippi, canna species have been widely used in smaller-scale constructed wetlands at mobile home and residential sites to provide a colorful, semitropical landscape effect. Because they are not cold hardy plants, however, they are not suitable for cold winter regions.

Cattail and reed have both been viewed by wetland managers, farmers, and others as wetland "weed" plants due to their propensity to outcompete species that might be seen as more desirable from the standpoint of waterfowl food or habitat, and both species have been quick to invade and sometimes take over areas with high water tables where standing water is never observed. Bulrush has the ability to grow in deeper water than cattail and will persist even under unusual flooding conditions; in addition, it is not as aggressive as cattail and is less of a threat to any existing adjacent natural wetland communities. Ornamentally, bulrush may be considered more aesthetically pleasing than cattail.

Although both terrestrial and emergent aquatic plants obtain carbon from the atmosphere in the form of free carbon dioxide, which enters the stomata on the leaves, many emergent aquatic plants possess unique adaptations for transporting atmospheric gases. One of these adaptations, previously mentioned, is represented by the intriguing and significant features known as *aerenchyma,* interconnected

Figure 2.6 *Common reed (Craig Campbell).*

tubes or gas-filled spaces that permeate most submerged stems, petioles, and leaf blades. These spaces allow oxygen and other gases to flow freely and be transported from the surface areas down into the root zone. It has been proven that continuous, uninterrupted tubular spaces can extend from the tip of a leaf, through the petioles and stems, and down into the buried roots and rhizomes within an anaerobic zone. This movement of gases, or ventilation system, involves pumping air against a pressure gradient and is produced by the heat of the sun, not by photosynthesis; it is a key characteristic that makes such plants valuable in renovating wastewater. It was shown that a severed water lily leaf will continue to draw in oxygen from the atmosphere and force it out the severed end of the petiole below the water surface, an activity that appears to halt when the leaf is shaded (Dacey, 1980).

Although the ability of many emergent species to transport oxygen to the root zone provides the basis for predicting high rates of denitrification, some wetland specialists have argued that there is evidence that the oxygenation primarily affects the rooting tissue, not the sediments, and that oxygen entering the sediments is usually instantly consumed microbially (Wetzel, 1993). The actual benefits, in terms of wastewater and stormwater renovation, that are directly attributable to an oxygenated root zone are not well established, and are in need of continued study and monitoring. As mentioned previously, various pilot experiments to date tend to support the view that there is a direct connection between the presence of roots and oxygen in the substrate and the enhancement of ammonia removal by the presence of roots rather than just soil or gravel alone. The problem in many past constructed wetlands would appear to be due to overemphasis on *Typha,* which often has resulted in a high percentage of the entire flow of wastewater passing below the natural root zone, and not in contact with the oxygenated zone.

Many researchers have demonstrated experimentally that aquatic plants are capable of concentrating, absorbing, and translocating heavy metals and even some radioactive elements. Some aquatic plants are able to absorb certain organic molecules intact, translocate them, and metabolize them with enzymes, rendering even some insecticides innocuous.

Submerged plants not in contact with the atmosphere must obtain their carbon dioxide directly from the water. When carbon dioxide is dissolved in water, it is frequently stored in the form of bicarbonates that many submerged plants can assimilate and decompose into carbonates, water, and carbon dioxide. Limestone in a stream channel or basin provides a source of carbon dioxide for photosynthesis by submerged plants, and hard waters are often more productive than soft waters. This is one reason that people with decorative ponds in a hard water area experience few problems with plant growth even though the evaporation and makeup water tends to create even harder water. In the author's experience, a decorative pond with a pH of up to 9 still supports both fish and lush aquatic plant growth, as the plants are capable of continually moderating the water's hardness by converting bicarbonates to CO_2, which is then utilized. This process, however, tends to involve precipitation of carbonate on the surfaces of the leaves and stems, giving them a limey texture.

The adaptability of some species of aquatic plants is remarkable. Not only can many of them survive when their normally wet habitats dry out, but some species are also capable of producing "land forms" that replace the normally submerged foliage during dry periods. Water milfoil, for example, is capable of producing new shoots with an entirely different form of growth when it is stranded by declining water levels. Many species of sedge (*Carex* spp.), while not changing in growth form, nevertheless are capable of withstanding conditions varying from complete inundation to total drying out for varying periods of time and are thus highly valuable for stormwater wetlands. Sedges have a wide natural distribution, and many species are now being propagated by native and aquatic plant nurseries. This group of plants offers great potential for creating attractive natural edges to wetland installations in areas near the interface between aquatic and terrestrial plantings; many *Carex* species are grasslike and attractive year round.

Until fairly recently, there were only a few sources of wetland plants in the quantities typically required for constructed wetland plantings. These suppliers were located primarily in Wisconsin, and their main objective was to encourage the development and enhancement of waterfowl nesting habitats in the Midwest. These nurseries were responsible for some excellent work with farmers and other land owners who wished to increase the suitability of their ponds, lakes, or marshes for waterfowl habitats. Many exotic plant species were introduced and are still propagated by these nurseries for their outstanding food value to waterfowl. Examples of such plants are Japanese millet (*Echinochloa crusgalli*), sesbania/swamp peas (*Sesbania macrocarpa*), and duckwheat/tartary buckwheat (*Fagopyrum tataricum*). Today, due in part to the increased interest in wetlands and the corresponding

Figure 2.7 Aquatic plant propagation greenhouse (Craig Campbell).

higher demand for wetland plants, both for restoration and for new constructed wetlands, the number of growers and suppliers of aquatic plant species has multiplied rapidly in all parts of the country, including the arid West. In addition to propagating the more common emergent aquatic plant species, many of these growers will collect seed and propagate any aquatic plant for custom orders if there is a need for particular plants in large quantities. Propagation activitites are typically undertaken in greenhouses, with container production outdoors during the growing season (Figures 2.7 and 2.8). Some of the wetlands nurseries, such as Green Acres

Figure 2.8 Outdoor aquatic container plant beds (Craig Campbell).

and Aquatic & Wetland nurseries in Colorado, have also installed constructed wetland systems to treat the wastewater generated by the nursery operations (Figure 2.9).

Payne (1992) reports that along with *Typha, Sagittaria, and Sparganium* species, a number of emergent *Polygonum* species are tolerant of pollution, turbidity, and related factors. The author is not aware of any experiments undertaken specifically with any of the *Polygonum* species to determine their potential usefulness in constructed wetlands, but due to the widespread natural occurrence of these plants throughout the country in widely varying climates, they should definitely be considered for pilot-scale wetland plantings.

There is often a need, particularly in arid zones, for plants that will withstand both inundation and relatively dry conditions. This requirement is particularly important for stormwater wetlands, but also for infiltration swales and basins. One of the grasses that has proven capable of withstanding wide variations in soil moisture is reed canarygrass (*Phalaris arundinacea*), which has been used to vegetate and stabilize marsh dikes due to its good root structure; its seeds are also a good wildlife food source. Another good plant for biofiltration swales and other areas exposed to varying moisture conditions, particularly in the West, is the Nebraska sedge (*Carex nebrascensis*), which is alkali tolerant, strongly rhizomatous, widely distributed, and an excellent soil stabilizer.

Mudflats are common edges in many artificial as well as natural wetlands and are typically colonized by plants such as barnyard grass, panic grass, bulrush, squarestem spikerush, smartweeds, millets, nutsedges, and other plants that often provide the most concentrated and attractive food sources for waterfowl. These border zones, which are subject to varying levels of moisture, are the most difficult to replicate successfully in a constructed wetland due to the typical controlled conditions created by waterproof liners. They are more likely to be a component of a stormwater wetland treatment system than a wastewater constructed wetland.

Figure 2.9 *Constructed wetland runoff treatment pond, Green Acres Nursery (Craig Campbell).*

PLANTING TECHNIQUES

Although seeding of wetland plants is often done where the timing of establishment is not critical, the normal method of establishing plants in constructed wetlands is by transplanting roots, rhizonmes, or tubers with some part of the abovewater stem attached. Some specialists feel that collected plants from existing wild stock will be better adapted to the conditions of a constructed wetland than nursery-grown stock. This is possible only in certain parts of the country and generally is a more costly method of obtaining plants than is the nursery-source method. In addition, it may be more difficult to control the timing and the precise quality of the collection operation when attempting to transplant from an existing wild source. After collection, wetland plants should ideally be planted as soon as possible, or within thirty-six hours of collection. This is often difficult, especially when the plantings are the responsibility of a subcontractor whose calendar may not coincide with that of the general contractor. Within the construction industry, the multitude of problems that can develop between subcontractors and a prime contractor are notorious and often focus on timely performance or the readiness of a site for a particular form of work. In the author's experience, plants collected from areas nearest the constructed wetland site and installed with at least 6 inches of cut stem still attached to the roots provide better growth than nursery plants brought in from a long distance. There is probably both a genetic adaptation and a time factor that favors the collected plants, but there is also the advantage of installing a relatively mature rootstock. Due to the carbohydrate stored in a mature rootstock, such plantings exhibit much greater growth during the first year of planting than does a tubeling plant that has just begun to establish a root system.

In general, container-grown plants have higher survival rates than bare root plants, even though their initial cost may be higher. In addition, there is less urgency to plant container plants immediately, and they can be stored or held for periods of time well beyond the time when bare root plants can be held. Even when the costs of container-grown plants are double those of bare root plants, the survival rates can offset the difference and result in lower overall costs. If, however, there are means of controlling precisely the timing for digging and transplanting more mature plants, this method can be quite effective. A number of nurseries specializing in wetland plants that are able to provide plants propagated from seed and sold in tubes or flats also have their own wetlands from which they regularly harvest mature plants. It has been demonstrated that many aquatic plants, such as *Phragmites,* exhibit much greater belowground than aboveground biomass. Because they are perennial plants, this could explain the advantages in terms of rapid establishment of planting rootstocks rather than seedlings. In many situations, however, precise control of plant delivery and planting operations is not possible, and nursery seedlings are the best alternative (Figures 2.10 to 2.13).

HISTORY OF CONSTRUCTED WETLANDS

The first scientific research studies and pilot-scale constructed wetland wastewater treatment facilities originated in Germany at the Max Planck Institute, where Kathe Seidel undertook detailed testing of many aquatic plants to determine their ability

Figure 2.10 *Planting bareroot* Eleocharis acicularis *in a tailings pond, Washington State (Bitteroot Native Growers).*

to absorb and break down chemical pollutants. Her research, first presented in 1953, proved that particular plant species such as *Scirpus lacustris*, a bulrush, had the ability to remove phenols, pathogenic bacteria, and other pollutants; in addition, plants grown in wastewater exhibited surprisingly varied physiological and morphological changes that aided their performance.

Robert Kadlec, principal of Wetland Management Services, is a distinguished researcher in the area of constructed wetlands who developed a full-scale experimental wastewater treatment system in 1973, while at the Institute of Water Research at the University of Michigan, that included a lake and a marsh system of 186 acres. Three 1.0-acre marshes were constructed to allow for flow from one of the lakes through the marshes and into another lake. The marshes were constructed in a terrace design that provided three zone depths of 18, 24, and 36 inches, allowing for the establishment of biota comparable with those of the natural marshes of the area (Bahr, 1974).

In 1975, Edward Furia, an attorney and city planner in Philadelphia, and Joachim Tourbier invited Kathe Seidel to consult on the design of a wetland polishing system for wastewater. They followed up by organizing the first international conference on biological wastewater treatment alternatives, which was held at the University

Figure 2.11 *Bareroot cattail with mature rootstock ready for planting. (Bitteroot Native Growers).*

of Pennsylvania in 1975. The proceedings, which were published under the title *Biological Control of Water Pollution* by the University of Pennsylvania Press, represents the first serious reference volume on the topic.

From 1976 on, research on the subject grew rapidly. Professors W. E. Sloey, C. W. Fetter, and F. L. Spangler of the University of Wisconsin-Oshkosh developed pilot-scale facilities, which resulted in their reports on the potential to replace septic tank drain fields by artificial marsh treatment systems (Sloey, et al., 1978).

Questions frequently are raised regarding the ability of constructed wetlands to function in colder climates. Some of the best data on that topic came from an extensive experimental facility constructed at Listowel, Ontario, in 1979 that was monitored between 1980 and 1984. Five separate marsh systems, or cells, which varied in size, configuration, depth, loading rates, and detention times, were tested with both primary and secondary effluent inputs on a year-round basis in a cold climate. Ice typically formed in winter on the surface of the marsh to a depth of approximately 4 inches, with wastewater flowing continuously below the ice all winter. The water depth was deliberately increased during the winter months to at least 12 inches to allow for a minimum of 8 inches of water below the ice, which also served to increase the detention time. Cattails, the only aquatic species used in the system, were discovered not to be totally dormant during the winter, and the

Figure 2.12 Planting Scirpus acutus *in hay bales at a pond edge, Washington State (Bitteroot Native Growers).*

Figure 2.13 Planting container plants in a mitigation wetland, Keystone Resort (Craig Campbell).

system provided acceptable removal mechanisms, although at a decreased rate, even in the coldest part of the year (Wile, et al., 1985).

Results from Listowel suggested that hydraulic loading rates of 200 m^3/ha/day (21,400 gal/acre/day) with detention times of about seven days provided maximum treatment efficiency. A narrow channel configuration proved to be the most effective design, and it was suggested that a length-to-width ratio of at least 10:1 be used (Reed et al., 1984). However, experience in other areas with subsurface flow systems has resulted in other recommendations for the length-to-width ratio to reduce the potential for clogging of narrow inlet cross sections.

As a result of the experience at Listowel, a full-scale constructed wetlands treatment facility was established at Port Perry, Ontario, which became operational in May 1986. The purpose of this facility, which involved retrofitting of former stabilization ponds, was to test the performance and operational requirements of a full-scale marsh treatment system and to assess the chosen method of pretreatment, which was a partial-mix facultative aeration lagoon (7.4 ac.) with 30-day retention time plus phosphorus reduction by continuous alum injection. The system was designed to treat all the domestic sewage from a community of 4,000 persons with an average annual flow of 0.4 mgd. Two separately monitored marsh systems, operated in parallel, were installed. System 1 was a wide channel marsh of 4.7 ac. with a length-to-width ratio of 13:1 that had complete cattail cover; system 2 was a rectangular marsh of 9.6 ac. with 30 percent cattail cover (Herskowitz, 1986).

In the mid-1980s the Tennessee Valley Authority (TVA), under the leadership of Dr. Donald Hammer, Senior Wetlands Ecologist, took the lead nationally in advancing the knowledge of constructed wetlands through its program of initiating a series of pilot-scale and full-scale demonstration projects that were constructed to treat the wastes from small towns in Kentucky with flows ranging from 100,000 to 500,000 gpd (0.5 mgd). Drawing upon their earlier successful experience in testing constructed wetlands to treat acid seepage from coal slurry impoundments in Alabama, the TVA then undertook a series of demonstration projects in cooperation with other agencies and has published, with updates, an excellent regional manual for homeowners and others interested in small-scale constructed wetlands systems. Dr. Hammer was instrumental in sponsoring the first major international conference on constructed wetlands in Chattanooga, Tennessee, in 1986; the proceedings are still considered one of the major reference books on the subject (Hammer, 1989).

Unfortunately, around 1993, the TVA greatly reduced its efforts within the field of constructed wetlands, leading to the departure of Dr. Donald A. Hammer, Gerald Steiner, and other key staff who had been centrally involved in constructed wetland research within the TVA. To date, no other agency at any level of government has conducted research, published reports, conducted conferences, and supported both pilot-scale and full-scale constructed wetlands at the level established by the TVA.

As efforts continued around the country, and as major conferences on the subject of constructed wetlands began taking place every few years, research and applications of constructed wetlands were supported at modest levels by agencies such as several regional headquarters of the U.S. EPA, the Soil Conservation Service, the Washington State Department of Ecology, the Maryland Department of Natural Resources, the Florida Department of Environmental Regulation, and others. These policies and projects helped to address the potential for constructed wetlands to treat urban stormwater runoff, wastewater, point source pollution, parking area pollution from transit vehicles, livestock wastewater, pulp and paper mill wastewater, and other sources of pollution. A separate line of investigation and projects developed the science of constructed wetlands to treat acid mine drainage, particularly in the state of Pennsylvania. This particular application of constructed wetlands has generated its own literature and is not addressed in detail in this volume. In 1980, two notable engineers in the field of wastewater treatment technology, George Tchobanoglous and Gordon Culp undertook one of the first in-depth engineering assessments of constructed wetlands for wastewater treatment and identified the need for additional research to quantify or refine the following interrelated factors:

1. Effect of plant type and biomass on degree of treatment achieved.
2. Effect of plant harvesting on nutrient uptake and degree of treatment.
3. Effect of bottom substrate on plant uptake and degree of treatment.
4. Effect of detention time on degree of treatment.
5. Effect of seasonal conditions on degree of treatment.
6. Effect of humus and litter components on degree of treatment.
7. Definition of removal kinetics as a function of plant type, biomass, detention time, and temperature.
8. Effect of wetland configuration on degree of treatment.
9. Definition of steady-state constituent removal capacity and constituent holding capacity as a function of detritus accumulation (Tchobanoglous G. and G. Culp, 1980).

It is interesting to note that while much more information has now been assembled on several of these issues—particularly those of detention time and configuration—the rest of the issues are precisely those still needing further attention today. While it is true that there are also more data and more understanding at this time regarding the effects of particular plant types, bottom substrate, detention time, configuration, and so on, most of this has been localized and not generated by regional pilot facilities in various climatic zones with complete control and comparable levels of monitoring. Rather than being the result of an organized, funded effort by the U.S. EPA on a region-by-region basis to develop regional guidelines based upon the widely varying conditions in different parts of the country, the data have instead been derived basically from reports by engineers or municipalities with different methods of monitoring, different loading rates, and different climates, with questionable means of correlating one set of data to another.

Engineering firms undertaking the design of constructed wetlands over the past ten years have done so on the basis of relatively incomplete regional information. This illustrates the problem that the U.S. EPA had intended to address properly beginning in the late 1970s, when their own reports identified the need for regional experiments at a pilot scale to determine the effects that variations in winter climate, altitude, loading rates, and so on might have on the design criteria. These experiments, if properly conducted and monitored, would have given the entire field of constructed wetlands with a much more solid database, with stringent and uniform methods of sampling and interpretation providing a respectable and supportable basis for design anywhere in the continental United States. At the time this effort was originally proposed, grant funds for wastewater treatment facilities—as opposed to loan funds—were still available to communities, and replacement money for failed systems was available to communities progressive enough to attempt to utilize experimental or innovative technology. Under this program, several constructed wetland treatment systems were originally designed as controlled units located within existing natural wetlands in the Pacific Northwest in Cannon Beach, Oregon, and Black Diamond, Washington, by the firm Kramer Chin and Mayo. Unfortunately, opposition by several federal agencies forced revisions to the original designs that resulted in less control and lack of adequate performance in the case of Black Diamond (Campbell, 1986). Unfortunately, the major cutbacks in EPA activities in the early 1980s brought to a halt much of the experimental work in constructed wetlands conducted or sponsored by the various labs and regional facilities of the EPA, leaving many unanswered questions that remain to this day.

One recent effort to assess the status of constructed wetlands for wastewater treatment with EPA sponsorship was led by Sherwood C. Reed, P.E., of Environmental Engineering Consultants and focused specifically on subsurface flow systems. A group of national experts in the field provided reviews of the draft report and participated in a two-day workshop in New Orleans, providing the basis for the final report of July 1993 (Reed, 1993), which represented a consensus on design, construction, operation, and maintenance of subsurface flow wetlands. This report identified a number of areas needing further research, which indicates the lack of funded and nationally organized research and monitoring efforts identified in assessment reports almost a decade earlier. The high-priority research needs identified in Reed's report are as follows:

- A better understanding of the nitrogen removal and nitrogen transformations occurring in these SF [subsurface flow] systems is necessary. This should lead to the development of national temperature and possibly seasonally dependent design models for nitrogen removal.
- Additional data collection is necessary on the spatial responses to BOD within the SF wetland bed to permit development and validation of improved design models to replace the interim procedures now in use.
- Further research is needed on identifying the oxygen needs and sources in these SF systems. The role of the plant roots in providing this oxygen is especially important.
- The use of plant types other than reeds, rushes, and cattails needs to be investigated to determine if optimum species exists. The need for routing plant harvest also deserves study.
- Most operational SF wetlands demonstrating successful ammonia removal have an HRT [hydraulic residence time] of about six days or more. The system at Benton, [Louisiana], has apparently demonstrated high ammonia removals with an HRT of <1 day. The factors responsible for this performance at Benton need to be defined.
- Although recent studies indicate minimal clogging in the beds investigated, the effort needs to be continued to determine the long-term risks of clogging.
- Efforts should continue to collect reliable performance data from full-scale operating systems to confirm and supplement the results from laboratory, greenhouse, and pilot-scale research.

An additional eight areas of research labeled as medium-priority and low-priority needs were also identified by the panel of experts contributing to this effort.

The executive summary of the report includes the following statement:

> This report verifies that SF constructed wetlands can be a reliable and cost-effective treatment method for a variety of wastewaters. These have included domestic, municipal, and industrial wastewaters as well as landfill leachates. Applications range from single family dwellings, parks, schools, and other public facilities to municipalities and industries. It can be a low-cost, low-energy process requiring minimal operational attention. As such the concept is particularly well suited for small to moderate sized facilities where suitable land may be available at a reasonable cost. Significant advantages include lack of odors, lack of mosquitoes and other insect vectors, and minimal risk of public exposure and contact with the water in the system.

It is worthwhile to list the conclusions of this technology assessment of subsurface flow constructed wetlands, as these are the types most commonly employed at present. The conclusions represent the current state of the art.

1. The subsurface flow constructed wetland concept offers high performance levels for BOD_5 and total suspended solids (TSS) at relatively low costs for construction, operation, and maintenance. It is particularly well suited for small to moderate-sized installations where suitable land and media are available at a reasonable cost.
2. The odor and vector control offered by the subsurface flow concept make it attractive for systems located in close proximity to the public. These uses range from single-family dwellings to larger developments and public facilities.
3. The cost effectiveness of SF wetland systems compared to free water surface (FWS) wetlands for the same water quality goals will depend on the local availability of land and the cost for land and for the media used in the SF concept.
4. Ammonia removal in most of the present generation of operating subsurface flow systems is deficient. The reason is believed to be the lack of oxygen in the bed profile and a too brief HRT to complete the nitrification reactions.
5. Effective ammonia removal has been reliably established in a few subsurface flow wetland systems. The common elements in those systems are full penetration of the plant roots and an HRT exceeding three days.
6. Removal of BOD_5 and TSS is not related to the length-to-width ratio of the system.
7. Surface flow has been observed in a number of operational subsurface flow systems. This is believed to be largely due to inadequate hydraulic design of the systems, not to clogging of the pore spaces in the media. The water level in the bed can be effectively controlled with adjustable outlet structures.
8. It is likely that some oxygen is available from the plant roots to support nitrification reactions. Effective use of that oxygen source requires complete development of the root zone in the bed profile and sufficient detention time. Neither condition is present in most operational SF systems. Further research is necessary to optimize these relationships.
9. Methods appear to be available to induce and maintain root penetration in order to enhance this oxygen source for nitrification. Approximately a six-

day HRT would be necessary for significant nitrification with a fully developed root zone and warm weather conditions. This approach cannot be used as a retrofit for most existing systems since there is not enough area available to increase the HRT to six days.

10. Use of a recirculating nitrification filter in combination with the subsurface flow wetland bed seems to offer promise for successful ammonia control and continued high levels of BOD_5 and TSS removal. This combination may be more cost effective than an FWS wetland designed for the same performance level.
11. The removal of BOD_5 in SF wetlands shows a linear relationship to the BOD_5 mass loading up to levels of at least 125 lb/acre/day.
12. A first-order plug flow kinetic model provides a reasonably accurate estimate of BOD_5 removal performance and is recommended for use as an interim approach until more sophisticated models can be developed and validated.
13. Darcy's law provides a reasonable approximation of the hydraulic performance in these SF systems as long as the limitations are recognized and accommodated.
14. The hydraulic conductivity (ks) and porosity (n) of the media to be used in these systems should be tested in the field or laboratory prior to final design.
15. To provide an adequate safety factor, no more than one-third of the measured "effective" hydraulic conductivity and no more than 10 percent of the maximum potential hydraulic gradient should be used for the hydraulic design of these systems.
16. SF systems of all sizes should include a final adjustable outlet to permit control of the water level in the bed.
17. Larger systems (Q = 5,000 gallons per day) should consider the use of multiple wetland cells in parallel to improve control and flexibility in operations.
18. The limited data available on removal of fecal coliforms indicate that these systems are capable of about a one- or two-log reduction with sufficient HRT. In most cases that will not be sufficient to reach the commonly applied limit of 200/100 ml, so some form of final disinfection may be necessary.
19. Some form of preliminary treatment, at least to the primary level, is typically used for SF wetland systems in both the United States and Europe. Primary treatment using septic or Imhoff tanks is suitable for small to moderate sized systems. Many existing SF systems follow facultative lagoons since they were typically added as a polishing step. Facultative lagoons are an acceptable form of preliminary treatment but can add large concentrations of algae. In these cases, a variable level draw-off in the lagoon may help reduce the algal load on the wetland component (Reed, 1993, pp. 9-1 to 9-3).

Another noteworthy pioneer in the field of constructed wetland research is Billy Wolverton, formerly with NASA's research facility in Bay St. Louis, Mississippi, who was responsible for much of the research on the ability of particular plants to take up various nutrients, chemicals, heavy metals, and other pollutants through the work he conducted for NASA in the 1970s. It was also Wolverton who first announced that certain plants typically used for indoor plantings, such as philodendron, had the ability to remove airborne contaminants such as formaldehyde. This led to considerable research by others later, and their results generally supported Wolvertons's conclusions. In a wonderful and well-illustrated recent book, Wolverton (1996) identifies the best indoor plants for removal of ammonia, formaldehyde, and benzene, with charts illustrating the relative effectiveness of each of fifty plants (Wolverton, 1996).

It was also Dr. Wolverton's work that inspired the use of constructed wetlands inside the Biosphere 2 experimental habitat near Oracle, Arizona, as described in the previous chapter.

CURRENT STATUS OF CONSTRUCTED WETLANDS

At this time, a great deal of the pioneering work by Billy Wolverton for NASA in Mississippi; by Robert Gearheart in Arcata, California; by Sherwood Reed nationally; by Dr. Don Hammer for the TVA; and by others made it possible, through their research efforts and pilot-scale experiments, for engineering firms such as Southwest Wetlands Group, Inc., Ron Crites and Nolte and Associates, CH2MHill, PBS&J, and others to take the next step by convincing regulators on a state-by-state basis that constructed wetlands were worthy of consideration for wastewater treatment in many different applications. Southwest Wetlands Group led the way in educating the regulators with the respective health and environmental departments with jurisdictions over wastewater in the states of New Mexico, Arizona, Colorado, Wyoming, Montana, Indiana, and Texas; Ron Crites of Nolte & Associates pioneered the effort in California; PBSJ and CH2Mhill took on the task in Florida and the Southeast; and other firms also made considerable efforts to gain approval and the required permits for constructed wetlands in other states. At this point, constructed wetlands for wastewater treatment have been built or permitted in every region of the United States and in several provinces of Canada.

3

Constructed Wetlands and Wastewater Treatment Design

The world that we have made as a result of the level of thinking we have done thus far creates problems that we cannot solve at the same level as the level we created them at.

—Albert Einstein

WASTEWATER TREATMENT GOALS: A BRIEF LOOK AT CURRENT CONVENTIONAL APPROACHES AND THE ALTERNATIVES OFFERED BY CONSTRUCTED WETLANDS

To understand current practices, it is helpful to look at the history and training of sanitary engineers. Almost everyone is aware that the Romans built sewers (cloacas) for the collection of human waste and the effluvium of the Roman city. Even older civilizations, such as the Sumerian, had drains in the middle of paved streets (Durant, 1954). One of the earliest books on public works describes the aqueducts and sewers of Rome (Clemens, 1913). These early sewers usually led to a nearby river and relied on the river to carry off human and animal waste from the city. The sewers were often supplemented by human workers who collected human and animal waste in wagons and delivered it to nearby fields, where it was applied as a fertilizer to the crops.

These wagons were in fact the primary method of disposal of human and animal waste from the city. This was probably the primary method of treatment in European cities right through the Middle Ages. Until the major European cities reintroduced the sewer, these "honey wagons" provided an abundant source of fertilizer for nearby small market garden farmers. Indeed, until 1860, Paris was a net exporter of food because the "French intensive gardening" practice incorporated human waste as a fertilizer.

Clearly, the honey wagons left something to be desired, especially during the warmer months. From all written accounts, the stench was overpowering, and both men and women resorted to the use of perfumed handkerchiefs to attempt to ameliorate the smell. Disease was rampant during the later Middle Ages, and the frequency of epidemics increased as the populations grew. The means by which the great cholera epidemic of Europe in the late 1840s and early 1850s was transmitted was little understood until John Snow, an English doctor, traced its recurrence to a contaminated public well. This discovery is even more remarkable because it predated the discovery of bacteria.

Sanitation was poorly understood, and until the discoveries of Jenner, Koch, and Pasteur, many people had little or no conception of the relationship between human waste and many of the infectious diseases. The great worldwide epidemics of typhoid, cholera, and polio are all waterborne diseases that were and are the result of contaminated water supplies. Other diseases, such as hepatitis A that are endemic to certain regions of the world, especially the Mediterranean, are propagated through the contamination of groundwater supplies by human waste.

As civil engineers began to understand the relationship between contaminated water supplies and raw, untreated sewage, they started to design sewer systems for the collection and disposal of human waste. Initially, sewer collection systems sought only to collect human waste and the runoff from rainfall on city streets. During the late Middle Ages and until the invention of the flush toilet, it was common practice for human waste to be emptied into the streets, or into pit privies that seeped into the water table or overflowed into the courtyards and streets. The gutter served to carry off human excrement, washing water, and rainfall into sewers that eventually emptied into nearby rivers and streams. As the cities grew, these once relatively unpolluted streams and rivers became open sewers. With the exception of cities in North America, Western Europe, and Japan, this practice is still common in most parts of the world. As cities grew, the rivers became black with sewage.

Sometime during the 1600s, some ingenious citizen in Germany recognized that if the raw sewage was introduced into a big artificial pond, the natural processes of water purification could begin to work. These early pioneers in wastewater treatment recognized that much of the problem consisted of the solids that exist in sewage. All the debris from the city streets, as well as the organic solids, could be settled out by allowing the sewage to enter a pond where the velocity of the water was appreciably slowed. Water leaving these early settling ponds was clear and generally odor free. However, until the 1800s, the need for these settling ponds and additional wastewater treatment was not clearly understood until the medical profession made the connection between disease and contaminated water.

Much of what is now practiced in the United States as sanitary engineering, a subdiscipline of civil engineering, began with English and German engineers and sanitarians in the late 1800s. It became very obvious to public health officials in most of the European countries and the United States that some form of sewer system and wastewater treatment was necessary to protect the public health. Obviously, sewer systems were the first step, and then simple treatment processes were added. Initially, these treatment processes consisted only of sedimentation basins. These basins were designed to use gravity to cause solids to settle out of the wastewater. The water, without most of the solids, would then be discharged into the river or bay. These basins would be periodically drained and the solids removed for land application.

This initial form of wastewater treatment proved to be fairly successful. A simple examination of wastewater characteristics would show why this is so. Approximately 99.93 percent of medium strength sewage is water see Table 3.1. Only 0.022 percent is organic solids requiring treatment (biochemical oxygen demand). If the solids are settled, approximately 30 to 40 percent of the BOD is removed. The water discharged from these primary settling basins was usually clear.

Because gravity was such a potent element in treatment, and most of the early problems that engineers faced were with sewer design, most sanitary engineers were educated by teaching them how to move water containing solids from point A to B. Sanitary engineering courses focused on hydraulics, open channel flow, pressure and flow rates in piping of different materials, sedimention basins and settling ponds. As late as 1954, the standard text on sanitary engineering (Fair and Geyer, 1954) had approximately 20 pages on the biology of wastewater treatment. Every sanitary engineer's education was focused on the requirements of engineering complex sewer networks, calculating the flow of water in sewers as the result of storm events, and the physics of particles settling in water.

This does not mean that the academic profession was ignorant of the biological processes, but the great demand placed on the engineering profession to accommodate the needs of the rapidly growing cities in Europe and the United States required thousands of young men and occasionally women to design these very basic systems. Societal demand dictated the nature of the basic civil engineering education, as it still does today. The need during this period focused the educational process on some very basic concepts, and those who chose this field were generally required to focus on water supply engineering as well.

Early in the twentieth century, various investigators in England, Karl Imhoff in Germany, and one of Imhoff's students in the United States (Giesecke, 1938) began to investigate the natural processes that occur in the breakdown and treatment of organic solids and soluble compounds in wastewater. During the mid-nineteenth century, the British had developed the concept of biochemical oxygen demand (BOD), a measure of the amount of oxygen required to oxidize and thus stabilize the organic materials in sewage. It is also a measure of the pollution load on the receiving stream. Since most of the rivers and streams were receiving untreated sewage, this method was used to quantify the pollution load. Since the rivers and streams were being continuously reaerated as they flowed to the sea, the engineers could estimate the treatment potential of the receiving stream.

When the Water Pollution Control Act of 1972 was passed in the United States, large amounts of federal funding became available to municipalities to construct and/or upgrade their level of wastewater treatment. Funding was also significantly increased for research and development. In 1964, standard sanitary engineering

TABLE 3.1 Typical Composition of Untreated Domestic Wastewater (Tchbanoglous, 1993)

Contaminant	Unit	Weak	Medium	Strong
Total Solids	mg/L	350	720	1200
Dissolved Solids	mg/L	250	500	850
Suspended Solids	mg/L	100	220	350
Biochemical Oxygen Demand	mg/L	110	220	400
Nitrogen	mg/L	20	40	85

texts such as Fair and Geyer's *Elements of Water Supply and Wastewater Disposal* discussed sedimentation and flocculation technology, chemical treatment, filtration technology using sand and gravel, and activated sludge systems. There was no discussion of lagoons, wetlands, or overland flow systems, although all of these systems were in use at that time. By the early 1970s textbooks covered more of the biological processes, and by 1980, the U.S. Environmental Protection Agency (EPA) discussed all of the possible wastewater treatment options, including natural systems, in a publication which provided an evaluation of alternative and innovative technologies. In 1993, natural systems were finally included in Metcalf and Eddy's standard textbook on sanitary engineering.

The large construction grant program during the 1970s and 1980s encouraged the design and construction of large, complex wastewater treatment systems. Every large municipality was urged to raise the level of its water quality, and since the money was essentially free, engineering firms were not encouraged to develop energy-efficient, low-cost alternatives. Indeed, because of the way engineers are compensated, the exact opposite took place. Since engineers' fees are usually based on a construction cost fee curve, larger, more expensive projects resulted in more fees for the engineers. There were no incentives for cost cutting.

As a result of federal funding and the training of engineering professionals, what we provided for our larger communities were highly complex, computer-controlled wastewater facilities that were energy intensive. Designs relied on sophisticated machinery including pumps, blowers, and aerators. Structures were made of concrete and steel. Process control relied on sampling and control systems that monitored the physical and chemical parameters of the wastewater treatment process. Many municipal treatment facilities had specially designed computer control rooms complete with large-scale schematic process control boards. The blinking lights and associated dials and gauges were quite impressive, but without highly trained operators, these computer control systems were useless.

Many cities discovered that it took a very expensive, highly trained labor force and a lot of electricity to accomplish the design goals established by the engineers. This does not mean that the systems did not work well. On the contrary, these wastewater treatment facilities did an outstanding job. As examples, Denver's and Washington D.C.'s wastewater treatment facilities were capable of producing an effluent that met drinking water standards. Smaller cities like Santa Fe, New Mexico, were also provided with such facilities, but often in such cities, the public works department struggled to provide capable management and staff. The complexity of the mechanical and process control of these larger systems was mind-boggling. This was a reflection of the technological age in which they were developed, and in fact, much of the command and control philosophy assumed that one could do a good job of wastewater treatment without paying much attention to the biology. The design philosophy was in essence to settle the solids, grow as much bacteria in highly aerated basins as quickly as possible using the most efficient aerators, and then mechanically dry or cook the resultant sludge for eventual deposition in landfills or oceans.

This technology worked very well, and as long as electricity was going for 2 cents per kilowatt-hour, the subsidiary issues of sludge management, equipment maintenance, and operator training appeared manageable. But when the oil embargoes of the 1970s drove the price of electricity to 8 cents per kilowatt-hour, the economics of wastewater treatment became questionable.

The EPA began to seek alternatives to solve this problem with an innovative and alternative technology grant program. The intent of this program was to encourage the development of less costly technologies, especially for smaller and intermediate-size communities. As the microbiologists and ecologists began to have more input into the essential processes that take place in wastewater treatment, it became clear to almost all investigators in the late 1970s that natural systems, that is, the ecologies of the pond, the marsh, and the meadow, had some significant advantages that made wastewater treatment systems utilizing these ecologies very attractive. The advantages are as follows:

- Natural systems are self-maintaining, self-regulating, and self-organizing.
- Natural systems are solar powered and/or powered by the energy stored in the organic content of the wastewater.
- The complex ecologies of natural systems have the ability to treat and degrade toxic organic compounds and complex metals in biologically stable compounds.
- They are simple to build, and because they are self-maintaining and self-regulating, they are easy to operate.

Engineers had some experience with natural systems. Sewage lagoons or stabilization ponds are in essence applied pond ecologies. Land application of primary treated sewage took advantage of the complex ecologies of the meadow and the underlying soil. Up to the last thirty to forty years, these processes were not well understood, and most engineers designing such systems had rather pragmatic approaches. Both ecologies relied on plants (algae in the ponds and grasses in the overland flow systems), and most engineers understood that it took water and fertilizer to grow plants, which of course are the primary constituents of sewage.

In the late 1950s, research by Kathe Seidel (Seidel et al, 1976), a German biologist at the Max Planck Institute, led to the discovery that the ecology of the marsh could also be very effective in the treatment of industrial and municipal wastewater. As ecologists in the United States identified the biological process taking place in natural wetlands, researchers at the Brookhaven National Laboratory, National Aeronautics and Space Administration (NASA), and the Tennessee Valley Authority (TVA), as well as many universities, began to apply this knowledge to wastewater treatment. Although this process was identified as a treatment option in 1976 (Spangler) and 1980 EPA publications on innovative and alternative technologies, quantifiable design guidelines were not published until 1988 (U.S. EPA, 1988).

Engineers need design formulas that will predict, fairly accurately, the performance of the wastewater treatment process, and until 1988 the design rules were essentially those used for the design of sewage lagoons or land application of sewage. It was observed, however, that wetlands outperformed both of the other options. A 1988 conference hosted by the TVA in Chattanooga, Tennessee, was the second conference on constructed wetlands, and was a milestone in attendance and resulting cooperative efforts between biologists and engineers.

Conferences such as the Chattanooga conference were notable in that they represented a fundamentally different way of thinking about wastewater treatment processes. Engineers saw the limitations of relying on physical and chemical processes, and the wetlands biologists and ecologists saw that their knowledge could be applied in very practical ways to solve wastewater treatment problems efficiently. This was also a time when the ideas of natural processes, recycling, and reuse were highly favored politically, and there was widespread support for the basic concepts that were espoused by the practitioners of natural systems design.

However, it soon became clear that the use of natural systems, that is the ecologies of the pond, the marsh, and the meadow, would have to stand on their own merits. The use of natural systems would have to be able to do the job that conventional technology was accomplishing with less money, less energy, and less maintenance. It would not only have to perform as well, it would have to do a better job because the use of natural systems represented a fundamental change in the way engineers were taught to think about wastewater treatment problems. It required engineers, raised in the hard sciences of chemistry and physics, to deal with the biologists, ecologists, and the practitioners of the so-called soft sciences. The fuzzy, unpredictable world of bacteria and plants was a real challenge for people whose entire education and professional experience depended on the ability to predict treatment processes using mathematically based models. Engineers need accurate mathematical models capable of predicting performance, because the discharge permits issued by the States have limits that must be met.

The biologists were, however, learning to become much more precise in their statements about life processes, and ecologists like H. T. Odum (Odum, 1960), developed mathematical models that describe the natural processes that take place in wetlands. This, of course, delighted the engineers, who could begin to design natural systems using the principles developed by the ecologists and biologists.

The biochemistry of nitrification, denitrification, phosphorus recycling, organic decomposition, biodegradation of toxic chemicals and hydrocarbons, metal chelation, and precipitation could be quantified. The principles of process chemistry were combined with the biological processes to produce mathematical models describing the most basic biological processes that were occurring in natural systems.

The models are simple and reflect our limited understanding of natural systems. Perhaps more importantly they reflect the very change in thinking that Albert Einstein called for. Clearly, a revolution is taking place in the biological sciences, and it is slowly overflowing into the realms of sanitary engineering and wastewater treatment.

Almost all of the wastewater technology that is currently in use is based on ideas that are at least forty years old. As George Tchbanoglous pointed out in a Water Pollution Control Federation keynote address, our colleagues in the agricultural engineering field have improved food production threefold during this period, while we are essentially working with the same technology. Natural systems are potentially the crack in the edifice of current practices. They are like the seedling growing in the sidewalk, and as we begin to develop the tools, techniques, and understanding of how natural systems purify water, we will be able to solve environmental problems created by polluted water more successfully.

Part of the challenge requires that engineers learn to work more cooperatively with microbiologists, wetlands ecologists, landscape architects, parks and recreation planners, and wildlife biologists as part of a team. A major revolution is taking place in the biological sciences, and it will have major impacts on sanitary engineering. Most of the developments in biological sciences have been in medicine and agriculture, but it is very clear that these same techniques will impact the way we treat wastewater.

The use of natural systems is an immature and rapidly developing technology that is being applied to all sorts of wastewater treatment problems. The level of sophistication and understanding is probably the same as that of corn farmers in 1820 when they were harvesting 20 bushels per acre. We have only a limited understanding of the relationship of the thousands of species in a wetlands environment, but as with the corn farmers, the yield and efficiency of the system can be increased with improved strains of plants and bacteria, and the addition of the proper micronutrients. No doubt some young biologist will do for natural systems technology what Mendeelev did for the chemists. Conceptually, the designs will be able to organize and arrange appropriate plant and microbes communities and ecologies to utilize all of the chemical compounds in the wastewater. As the sanitary engineer becomes the ecological engineer, compounds in the wastewater will be viewed as resources to be converted to beneficial compounds, recycled to the atmosphere, or applied to the landscape. The conceptual change from viewing wastewater as a liquid with undesirable compounds to a liquid feed to a biochemical factory that removes all undesirable compounds, converting them to useful products and chemicals, requires a complete change in thinking. This paradigm shift is the challenge for our society. This is the time to bring together the various disciplines to reach for the unity of knowledge that Edward O. Wilson writes of in *Consilience* (Wilson, 1998).

CAPABILITIES AND LIMITATIONS OF CONSTRUCTED WETLANDS

In a very general sense, understanding the function of constructed wetlands requires that we step back in time to achieve the more basic understanding that every farmer had: plants require water and fertilizer for their growth. Sewage is essentially water and fertilizer. Wetlands plants require water and fertilizer, and unlike dryland plants, they can grow in saturated soils and standing water and consume several times the nutrients used by dryland crops. Thus the use of wetlands as a treatment process can be considered a form of agriculture, with the "crop" consisting of clean water.

Wetlands as an ecology are more productive in biomass production than any other environment except the rain forest. The prodigious amounts of biomass that are produced are the result of the equally large masses of nutrients and sediments that are washed into the lowlands where most wetlands occur. The very complex ecologies of microorganisms, plants, amphibians, fishes, mammals, and birds that make up the wetlands environment have evolved to such a level of efficiency that very little of the nutrients, organic solids, and sediments entering the wetlands leave. These very properties are the same properties that are desirable in wastewater treatment.

Wetlands are one of the principal ecosystems on the planet for recycling the essential elements of life (carbon, hydrogen, oxygen, nitrogen, and phosphorus), as well as the metallic micronutrients. They are the planet's kidneys, purifying the waters, and they have been doing so in varying forms for probably 250 million years. Our modern understanding of the importance of wetlands is relatively new and began with the sciences of biology, limnology, and ecology. There is, however, the biblical injunction in the Book of Moses instructing Moses to throw a particular stick into the water to remove the bitter taste. This is apparently the first recorded instance of ecological engineering. The ability of the wetlands ecology to remove metals (the bitter taste) has been well known by microbial ecologists for at least fifty years. The application of this knowledge to precipitate metals in acid mine drainage is relatively recent, beginning in the 1980s. Much of the early work was done by the Bureau of Mines in Pennsylvania (Bureau of Mines, 1988), and the TVA (Brodie, 1989).

Wetlands are generally used as one part of a multipart treatment system, and therefore their particular abilities can be employed to great benefit if the designer understands their essential abilities. Wetlands can be thought of as biological filters with certain basic abilities and limitations; each of the following sections describes both. Keep in mind that wetlands operate in conjunction with other treatment elements. The basic functions can be summarized as follows:

- Nutrient (nitrogen and phosphorus) removal and recycling
- Sedimentation
- Biological Oxygen Demand (organic compounds capable of being oxidized) digestion and removal
- Metals precipitation
- Pathogen removal
- Toxic compound degradation

Removal of organic compounds, both soluble and solid, that are typically measured by the biological oxygen demand (BOD) and by reporting the amount of total suspended solids (TSS) are removed by biological activity as well as by the process of sedimentation. The accumulation of organic sediments is the process by which peat and eventually coal are created. Biological processes return carbon to the atmosphere as methane and carbon dioxide or store large amounts of it in the biomass.

Since we recognize that wetlands are filters, it is not difficult to imagine that there are limits to the amount of organic and inorganic material that the filter can remove without encountering problems. Organic material does not disappear; it is converted to plant material, returned to the atmosphere, or deposited on the bottom of the wetlands, or it is discharged to downstream waters. The generally accepted range of carbon, defined as BOD that a wetlands can accept is expressed in terms of 112 kg/ha/year.[12] Exceeding this limit is likely to produce microbiological processes that are entirely anaerobic and may subsequently lead to the collapse of the ecosystem.

As Figure 3.1 indicates, the processes involved in recycling carbon are part of the natural cycle that is an essential element of the wetlands ecology. This recycling process is temperature dependent. Microbial activity can double as the temperature rises from 10°C to 20°C. Carbon is returned to the atmosphere much more quickly in the tropics than in the Arctic, which is why tropical rainforest soils have so little leaf litter relative to the biomass of the forest canopy, while peat bogs are common in the Arctic regions.

Most of the carbon remains in the wetlands in the form of highly reduced, long chain carbon compounds or in the form of peat-like plant materials. The retention

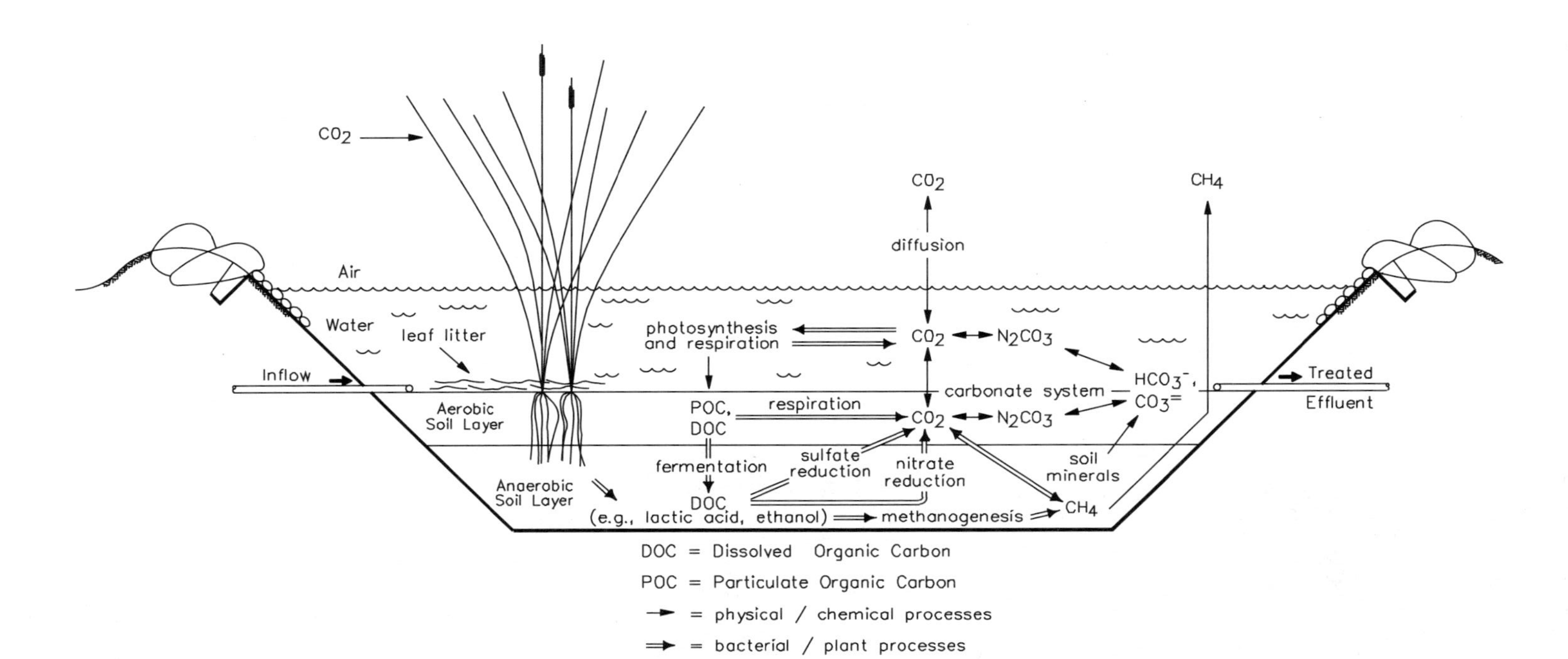

Figure 3.1 *Carbon cycle in wetlands.*

of carbon in wetlands is a significant part of the geologic cycle of carbon. In natural wetlands, these forms of carbon eventually become the future coal and oil deposits. Since one of the treatment goals is to remove carbon from wastewater (measured as BOD), passive treatment technologies such as wetlands should be considered because they offer significant savings in atmospheric carbon. Conventional, energy intensive, wastewater treatment plants discharge four pounds of carbon into the air, generating the electricity to remove one pound of carbon from the wastewater.

As the next chapter describes in more detail, the rate of removal of organic material is defined by the temperature of the water in the wetlands. However, this removal rate is further modified by the construction of the wetlands. Subsurface flow wetlands and surface flow wetlands have different physical characteristics and handle solids differently. Ideally, the uniform distribution of solids would clearly improve the efficiency of both types of wetlands. Typically, most problems with wetlands occur because the engineer did not properly design the front end of the wetlands to receive influent solids.

The removal or use of nutrients (nitrogen and phosphorus) by the wetlands ecosystem is a temperature-dependent process that is primarily biological, although it may also be chemical. The limitations are seasonal, and reflect the coincidental growth of plants and the rising temperature. Wetlands plants require more nitrogen and phosphorus in the growing season. However, the plant nutrients are subsequently released to the water during the fall and winter. This uptake of nutrients is misleading because the nitrogen and phosphorus are only temporarily removed. Phosphorus may be bound by clay particles in the soil, or transported away from the wetlands by insects and birds consuming seeds and leaves. Nitrogen removal occurs because of microbial activity recycling nitrogen to the atmosphere. Figure 3.2 shows the complex pathways for nitrogen in the wetlands environment.

THE PLANNING PROCESS

The design of new wastewater treatment facilities offers the engineer the opportunity to start from scratch. The process is comparable to that of designing a new house. The engineer has the opportunity to use the best available technology, to create the most efficient and cost-effective treatment facility, and to design a facility that may win awards. But of course, as with a house design, the main question becomes whether it is best for the engineer or best for the community being served.

Engineering firms are supposed to be knowledgeable about all aspects of the wastewater design process; they are expected to be aware of all the treatment options available. But, of course, that is impossible. Engineering firms, like individuals, have preferences. These preferences may be the result of education, experience, or regulatory agency policy. As an example of the last factor, many state environmental agencies determine what types of systems can be used for wastewater treatment. The State of Ohio, for example, does not permit the use of subsurface irrigation or subsurface infiltration systems for the treatment and disposal of wastewater. Slow-rate or rapid infiltration basins, described in detail in the U.S. EPA *Manual: Land Treatment of Muncipal Wastewater* (U.S. EPA, 1981) are not permitted in Ohio. As an example of design preference, the engineers of the Indian Health Service, a branch of the U.S. Public Health Service, recommend sewage

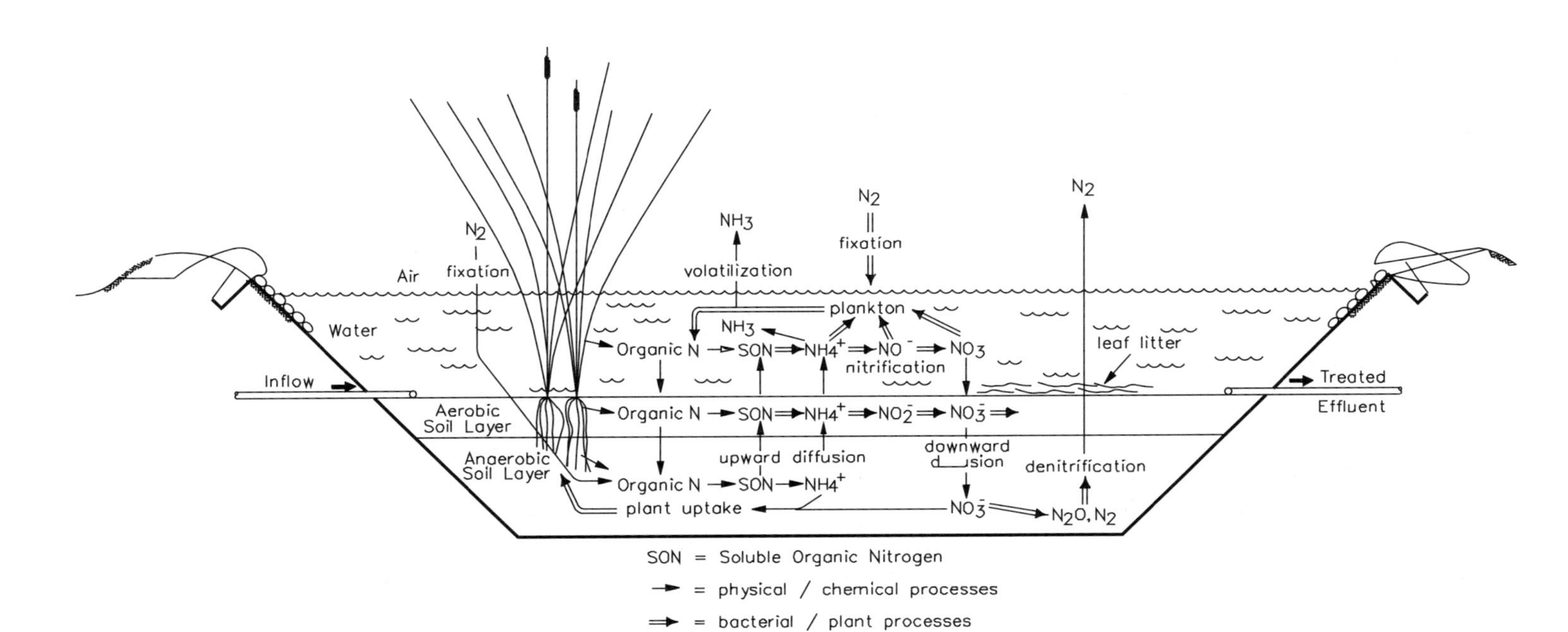

Figure 3.2 *Nitrogen cycle in wetlands.*

lagoons (developed in the 1920s) for use on Indian reservations. The result of both preference and regulation is that the clients being served do not necessarily get what is best for them. This is the tyranny of the expert.

To counteract this problem, the U.S. EPA devised some procedures that are usually very effective in developing a solution that is cost effective and meets the community's needs. The use of natural systems is often the best choice, but other alternatives should be considered and evaluated. The process of getting there is extremely important. The essence of the process is getting the community involved.

As an example of how this process can work most effectively, consider LaGrange County, a small rural county in northeastern Indiana. The Indiana toll road (I90) runs along its northern boundary. LaGrange County is a rural farming area, with a little industry in a region of lakes in the southeastern quadrant. The county planner recognized that to preserve the rural quality of life, a countywide land use plan would have to be developed. Because planning funds were limited, he enlisted the assistance of the Department of Architecture and Planning at Ball State University in Muncie, Indiana.

Because LaGrange County was in the likely path of development along the Indiana toll road, one of the planning recommendations was that the county create a countywide sanitation district to establish a uniform standard for development and to provide a mechanism for the future construction of wastewater treatment facilities and sewers. The small towns in the county did not have the resources to construct larger, more complex treatment facilities, and because of the rural nature of the county, it was clear that any new development would strain existing facilities or would be built in regions where there were no treatment facilities. More important, it was clear that the unincorporated areas needed some mechanism to control the water quality of the lakes and streams in the area.

Lake residents had previously complained of summertime algae blooms and bad odors which led the local county health officer to test the lakes to confirm contamination from septic tank leach fields. Agricultural runoff from feedlots and groundwater contamination from increasing numbers of septic tank leach fields were also identified as problems. As in most rural counties, the widespread use of septic tanks and leach fields, which were developed in 1871 for use by farm homesteads, had led to nitrate contamination of many wells in the county (Grant, 1996).

Problem recognition and definition is typically the first stage that a community must go through. Some farsighted members of the community may see the problem, but the next step is to bring the rest of the community into discussions, to consider potential courses of action, to document the problem, and to develop a plan acceptable to the majority of the community.

Once the sanitation district was established in LaGrange County, a board of directors was charged with the responsibility of developing a plan of action for dealing with the problems of population growth and its impacts on water quality. Thus, the second step is the development of a plan of action. Typically, this requires the planning group to establish a budget, develop a time line, document the nature and size of the problem, and begin the process of public involvement. This second phase of planning should include discussions of possible alternatives for wastewater treatment. The planning group should seek out and evaluate all of the possible options and, if necessary, engage consultants who are specialists in the proposed technologies.

Information gathering is an important part of the planning process, and various organizations have been developed to assist planners and the public in gaining information and developing a wastewater treatment plan. The U.S. EPA, the Water Environment Federation, the American Society of Civil Engineers, Small Flows Clearing House, and National Onsite Wastewater Reuse Association all have excellent publications written for the public that will assist in the process of developing alternatives. Since the number of alternatives may be overwhelming, it is often prudent for consultants to be engaged to help evaluate them.

The use of consultants to assist in planning should be a formal process. In LaGrange County, the sanitation district board of directors solicited assistance by advertising a Request for Proposals (RFP). Since the board had a fairly good idea of what they wanted after reading about the alternatives, the RFP clearly stated that they wanted a team of engineers to evaluate the alternatives under consideration. Eight teams of engineers responded, and based on their written submissions, four teams were interviewed, with a formalized evaluation scheme used to help pick the best team for LaGrange.

Once the selection was made, the board had to discuss the fees for the task. This negotiation was based on the scope of the task and the engineer's estimate of the number of man-hours that the evaluation of the wastewater treatment alternatives would require. Once the fees were agreed upon, the board then sought funding from various state agencies.

Most states have planning grants available to assist communities to develop wastewater treatment plans. The federal government encouraged the use of these grants and provided initial funding grants to the states in order to stimulate the planning process. The development of an affordable plan is typically the hardest step for any community for several major reasons. Many communities have learned, from the experience of other communities, that wastewater treatment and collection systems can be very expensive. As in LaGrange County, residents who previously paid nothing for wastewater treatment are now faced with a bill that may be $40 to $80 per month.

The engineering design firm that designs the wastewater treatment facility should be selected based on the following:

1. Experience with the technology proposed by the planning and feasibility study recommendations.
2. The best prior record with other similar communities, that is, previous community clients should regard the candidate favorably.
3. Staff adequate to complete a project of the size proposed.

There are many evaluation schemes available to assist the community in selecting the appropriate engineering firm (see appendix). Use of these evaluation schemes is highly recommended.

Only after the evaluation is made and a firm selected should fees be discussed. Based on the amount of work required, the engineering firm that is first on the list should be interviewed to discuss the level of effort required and the fees necessary to complete the project. Fee negotiation is very important and should be considered an essential part of final contract negotiations. Small communities may feel at a disadvantage here and should select appropriate counsel. Other professional engineers can assist in these negotiations.

The purpose of these negotiations is to reach an agreement that is satisfactory to both parties and to establish a clear understanding of the level of effort required, the exact nature of the work to be produced, review procedures, public meetings required, and all other aspects of a standard community project. The community should not select the engineer with the lowest fee, but rather the most qualified one. The consequences of poor design will affect the community for twenty or more years. Engineers' fees form an insignificant portion of the total cost of the project and, when added to a twenty-year life cycle cost of a conventional wastewater treatment facility, the engineers' fees could easily be less than 1 percent of the total capital and operating costs. For this reason, fees should not be the basis for decision making.

DESIGNING FOR NEW FACILITIES

Typically, the design of a wastewater treatment facility is the responsibility of the engineer. This responsibility is jealously protected by the civil engineering profession. Given the technical requirements, experience, and knowledge needed to design such facilities, this is not an unreasonable requirement. However, because of the facility's impact on the surrounding area, and the potential for reuse of the treated effluent for habitat development or landscapes, the design team should not be made up exclusively of engineers. Nevertheless, understanding the engineer's design process is essential for intelligent cooperation.

Keeping in mind the previous discussions concerning the pitfalls in designing a cost effective wastewater treatment system, the design team must still get on with the task. Assume that the design team has completed all of the tasks in the facilities planning guidelines, and that the feasibility study indicated that constructed wetlands are the most cost-effective choice. The following checklist contains the information required before proceeding with design:

- Estimate design flows.

  Determine current and future population (design flow).

  Determine how many people will be served during the life of the system. Multiply population by 40–60 gallons per capita per day to determine flow.

  Estimate wastewater flow from commercial and industrial facilities, if any. Use Uniform Plumbing Code for guidelines in the absence of specific information.
- Characterization of wastewater flow.

  Decide whether this is high strength wastewater or whether it has been diluted by high rates of infiltration of fresh water into the sewer system.
- Discharge permit requirements.

  Determine the regulatory requirements for surface discharge, land application, or subsurface disposal.
- Preliminary site evaluation.

  Visit possible treatment sites in the area to select the most suitable ones.

- Site topography.
  Obtain a 1 to 2 foot contour map of the site(s).
- Site soil and groundwater conditions.
  Determine the types of soils on the site, percolation rates, and depth to groundwater if an on-site discharge or land application is being considered.
- Climate.
  Evaluate precipitation and temperature in January and June, and determine soil temperature, pan evaporation, number of frost free days (growing season), and average solar radiation.
- Sewer system.
  Treatment and collection systems are tied together. For example, s conventional gravity systems might use aerated lagoons or large anaerobic pretreatment tanks. A small-diameter collection system might use on-site interceptor tanks for pretreatment.

Estimate Wastewater Flows

Many surveys have been made to determine the per capita wastewater flow. The U.S. EPA, in Table 6.2 of the *On-site Wastewater Treatment and Disposal Systems* design manual gives a range of 47–52 gallons per person per day. Individual states may have different values, which are generally higher. For example, Indiana uses 65 gallons per person per day, New Mexico 75 while Ohio sets 400 gallons per home per day. These values are clearly too high when compared to actual surveys of water usage. If they are used, the wastewater treatment system will be much larger than necessary. There may be similar systems or communities in the area, and a survey of existing systems may be acceptable to the regulator. Average flows per home can then be developed, which may range from 115 to 222 gallons per home per day.

After deciding which value is the most representative, the designer must then estimate the number of homes that will be built during the design life of the system. Typically, this will be 20 years. Often the past growth rates for the community or the county will be used, but this is obviously very problematical. Often communities with stable growth rates of 3–4 percent over the past 50 years will suddenly grow at the rate of 10–20 percent.

Characterize Wastewater Flow

The designer will need to know the expected BOD, TSS, total nitrogen, and total phosphorus. These values will depend on the nature of the community and the number of commercial, industrial, and restaurant facilities. If interceptor tanks are used, the pretreatment provided by these tanks will change the water quality significantly. Older sewer collection systems can also significantly dilute the wastewater due to infiltration of groundwater. However, this dilution will increase the flow, and therefore the hydraulic capacity requirements for the treatment facility.

If the collection system is in place, sampling of the wastewater for a week will provide an accurate measure of the wastewater water quality parameters. In the absence of these measurements, sources such as Metcalf & Eddy's *Wastewater*

Engineering (Tchbanoglous, 1991), or the U.S. EPA's manual *On-site Wastewater Treatment and Disposal Systems* (U.S. EPA, 1980) can be used to select appropriate design values for the required water quality parameters. Table 3.2 is typical:

Discharge Permit Requirements

Every state has its own set of water quality parameters for various discharge options. Those options typically include the following:

- Surface discharge into the waters of the United States.
- Land application
- Subsurface disposal/disposal to groundwater.

Surface discharge parameters will usually regulate BOD, TSS, nitrogen, and often phosphorus. The form of nitrogen that is usually regulated is ammonia, and often this value must be less than 2 mg/L. This will require some form of nitrification/denitrification (nitrogen removal) capability to be designed into the treatment process.

BOD values will usually start at 30 mg/L for discharges into large rivers, where the impact of the discharge is therefore small. Smaller streams usually require cleaner water, with 10 mg/L currently being the strictest standard in most states, although some states have limits as low as 5 mg/L. The difficulty and expense of reducing BOD values quality from 30 to 10 mg/L is not a linear relationship but rather an exponential one. Going from 30/30 to 10/10 can require a twentyfold increase in energy using conventional systems.

TSS present problems, especially in summertime with pond-based systems because the nutrients in wastewater encourage the growth of algae. Usually, the TSS values are arbitrarily set to be the same as the BOD values. Typically, permits are

TABLE 3.2 Characteristics of Typical Residential Wastewater

Description	Concentrations (mg/L)
Total solids	680–1000
Volatile solids	380–500
Suspended solids	200–290
Volatile suspend solids	150–240
Biological oxygen demand	200–290
Chemical oxygen demand	680–730
Total nitrogen	35–100
Ammonia	6–18
Nitrites and nitrates	<1
Total phosphorus	18–29
Phosphate	6–24
Total coliforms	10^{10}–10^{12}
Fecal coliforms	10^{8}–10^{10}

Source: U.S. EPA *Onsite Wastewater Treatment and Disposal Systems*, Cincinnati, OH, 1980.

usually expressed, for example, as 30/30, which means that BOD and TSS values must be less than 30 mg/L.

Land application permits are often more lenient than those for surface discharge systems because of the ability of crops to utilize ammonia and nitrates, and the ability of the soil to absorb and degrade organic material. Phosphorus is also utilized by crops. Every state has established values for crop uptake of nitrogen and phosphorus, and these can be significant. Often the problem for land application is the limited growing season, especially in the northern half of the United States. This requires that some form of storage be incorporated in the design. Figure 3.3 shows approximate water storage requirements in the United States.

Land application can also include reuse for irrigation of landscape. This option is very common in the West, and if a discharge is eliminated, the resulting savings in monitoring costs can be significant.

Subsurface disposal of wastewater is often the most economical alternative because water can be discharged all season long and because soil has some treatment capabilities. Soil can remove nitrogen, phosphorus, bacteria, and viruses. Many states define the water quality parameters for discharge into the groundwater in terms of drinking water parameters, or they specify that there shall be no net degradation of the aquifer. Each alternative must be evaluated to ensure that the most cost-effective option is selected.

Preliminary Site Evaluation

If a new wastewater treatment site must be selected, site maps that include the entire community being served can often be used in conjunction with soil conser-

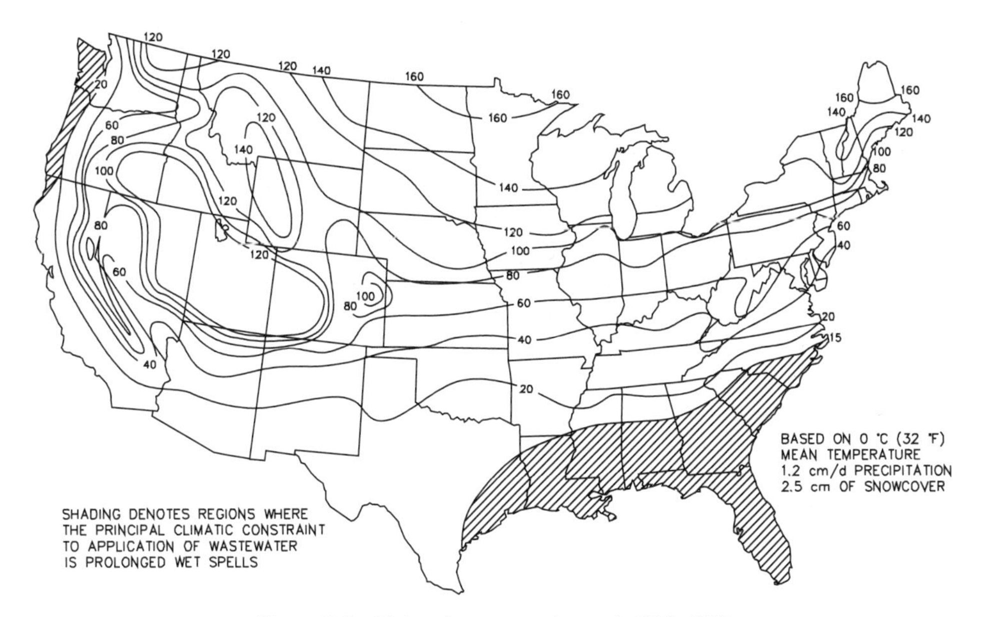

Figure 3.3 *Water storage requirements (U.S. EPA).*

vation maps to select appropriate sites for the wastewater treatment facility. The designer should then visit each site and rank the sites in terms of suitability.

Site Topography

After preliminary site selection, a topographic map should be obtained for the most likely site. This map should identify property boundaries, existing structures, wells, utilities, and other relevant structures such as field drain tiles and trees. The contours on this map should be 1- or 2-foot intervals. If surface discharge is being considered, the name and flow of the nearest stream or river should be indicated on the map.

Site Soil and Groundwater Information

Soil boring information to a depth of at least 60 inches at at least three sites should be obtained. More borings should be taken for larger systems or if the variability of the borings is large. Soils should be classified as to type and permeability. Permeability should be determined with the use of basin tests, not hole percolation tests. Basin tests should use basins of at least 75 square feet.

Depth to groundwater should also be noted. Water quality information should be obtained about the underlying aquifer if subsurface infiltration, infiltration basins, or land application are being considered for final treatment and disposal. Groundwater water quality parameters should also be determined. If possible, groundwater wells should be drilled to determine the hydraulic gradient and the nature of the aquifer.

Discharge to groundwater can provide significant additional treatment in the form of physical and chemical processes that provide additional filtration within the aquifer. Discharge into the subsurface aquifer is somewhat similar to discharge into a large, slow-flowing river or large lake. Water quality can be monitored by placing groundwater monitoring wells down gradient from the discharge area.

The treatment capabilities of soil and the associated physical, chemical, and biological processes have been well documented since the late 1940s by the U.S. Public Health Service. Subsequent investigations by many university researchers have provided additional information on the abilities of soil-based treatment systems. Since soil-based treatment offers significant cost savings over surface discharge in both operation and construction, this element of final treatment should be thoroughly investigated. Adding an additional treatment step with constructed wetlands is perhaps the single most cost-effective treatment option available to small communities.

Climate

All wastewater treatment facilities are ultimately affected by the water temperature. Natural systems like stabilization ponds, aerated lagoons, constructed wetlands, or land application systems are even more temperature dependent. It is important to obtain the following information:

- Monthly average temperatures, especially for January
- Precipitation, including 10-year 24-hour storm events
- Evapotranspiration
- Soil temperatures at 30 inches
- Solar radiation

Most of this information is available in Soil Conservation Service County reports or in National Oceanographic Aeronautics Administration (NOAA) state climate information reports.

Microclimates may have a significant effect on the performance of natural systems. South-facing slopes tend to be warmer than north-facing slopes. Systems located in valleys may be a lot colder in the winter than those placed on the sides of the valley. In the absence of climatic records, examination of local vegetation can be correlated to plant hardiness zones and subsequently to probable January minimum temperatures.

Sewer System

Existing sewers systems may have a significant impact on the volume and quality of wastewater being treated. Conventional sewers often accept storm runoff. The effects are increased and highly variable water flows and concentrations of pollutants. Older sewer systems often allow groundwater to infiltrate into the lines or surface water to flow into the manholes. Infiltration and inflow can be both seasonal and unpredictable. Annual records of seasonal flows are essential to determine the ultimate hydraulic capacity of the wastewater treatment facility.

If the community has a combined sewer/storm system or if inflow and infiltration are significant, it may be necessary to upgrade the sewer collection system as well. In any event, it is often necessary to spend some time investigating the sewer system.

New sewer collection systems generally do not have significant infiltration and inflow problems, and as a result, the volume of wastewater is generally far lower than in older conventional sewer systems. Because the per capita volume of water is less, the concentrations of pollutants are higher; there is less dilution from ground or surface water. Sometimes it is not only feasible but also cost effective to leave existing septic tanks in place and hook every one up to a small-diameter gravity or pressure system. The volume of wastewater is usually decreased along with the pollutant load.

Although a discussion of sewer collection systems is beyond the scope of this text, they are a significant part of the design process and should be considered very carefully prior to design. There are several excellent publications including *Alternative Sewer Design* published by the Water Environment Federation and *Small Diameter Collection Systems* by the U.S. EPA.

Design Examples

The following examples have been presented to show how the use of constructed wetlands has been incorporated into different small-community designs. When reviewing these examples, remember that constructed wetlands are one part of a multipart system. The first two examples are small communities with existing sewer systems that required entirely new wastewater treatment systems. The third example is of a small community that did not have a sewer system in place and required both new treatment facilities and a new collection system. The fourth design example is of a system in Mexico which treats sludge as well. Finally there are two examples of subdivisions: one in Indiana, and one in Vermont.

Ouray, Colorado The City of Ouray is located in the San Juan Mountains of southwestern Colorado at an elevation of 7,580 feet. Ouray is an old mining town situated in a narrow valley with the opening to the north. It is surrounded by mountain peaks ranging from 12,000 to 13,000 feet. The Uncomphrage River, which runs right through the town, drains an old mining region of Colorado. Consequently, the quality of the river is not very good.

Ouray had been experiencing a period of relatively rapid growth during the early 1980s and consequently had outgrown its existing wastewater treatment facilities. The old system was not worth salvaging and, like the facultative lagoons in the Village of Capitan (see following example), they had become filled with sludge. In addition, the mechanical aeration equipment needed replacement. Ouray had a sewer system, which, although experiencing a lot of infiltration due to high groundwater conditions, was still serviceable.

Because Ouray is located in a narrow canyon, the land available near the river was very limited, so no land treatment options were available. Options that were considered included aerated lagoons, activated sludge, and an oxidation ditch. During preliminary design evaluation, aerated lagoons were found to be the least expensive option and were the first choice. However, as with all lagoon systems, there is a high probability that during the summer months at some point the TSS will exceed the permit limits. The reason for this is that western waters are high in dissolved solids, and this, in conjunction with the high level of sunshine in the summer, causes water in lagoons to grow a lot of algae.

Since the aerated lagoons had already been designed, it was a simple matter to consider the effects on BOD and TSS removal in surface flow constructed wetlands. To save energy, the aerated lagoons would be designed to be operated with BOD and TSS levels of 60 and 75 mg/L, respectively, instead of 30/30, as the discharge permit required. The wetlands therefore had the function of removing the rest of the BOD and TSS down to the permit limit of 30/30.

Obviously, one concern was cold weather operations; temperatures in the Colorado Rockies at 7,500 feet are very low. Would the system freeze? How would it perform in the cold months? Because constructed wetlands had never been operated at this elevation and climate, provisions were made to bypass the wetlands if necessary. The aerators would simply have to be turned on more often during cold periods.

An important design consideration that was absent in this case is that there are no nitrogen limitations on the permit. The reason for this is that the Uncomphragre River's water quality is poor, so the addition of nitrogen would have no adverse affects. This obviously simplified the design problem. The design calculations (see below) indicated that the permit limits could be met with a 2.2-day detention time.

As you can see in Figure 3.4, the engineer decided to use six cells operating in two parallel sets of three. This allowed some operating flexibility and provided for additional mixing at the end of each of the first two cells in each parallel train. This is an important concept to understand. Consider the design for Ouray as built and then the alternative with one long channel. In the Ouray example, the water is mixed at the end of each cell, while in a single long cell, streaming effects can develop. This is similar to the situation in natural wetlands, where there is a clear channel in the center and vegetative growth is confined to the edges of the channel. If streaming develops, the detention time is shortened and the treatment is minimal.

Other techniques can help with the streaming problem. Introducing water at a single point is more likely to result in streaming effects because of the velocity of the water. Clearly, it is better to introduce the influent at all points at the head of each wetlands. Uniform distribution is an important part of the development of uniform stands of vegetation. As Figure 3.5 illustrates, the perforated pipe laid at right angles to the direction of flow will result in an almost uniform distribution of effluent.

Perhaps a more subtle point is that when the pipe is laid in the trench at the head of the cells and is underwater, hydraulic principles dictate that there will be equal flow from each hole, given a large-diameter pipe where frictional losses are

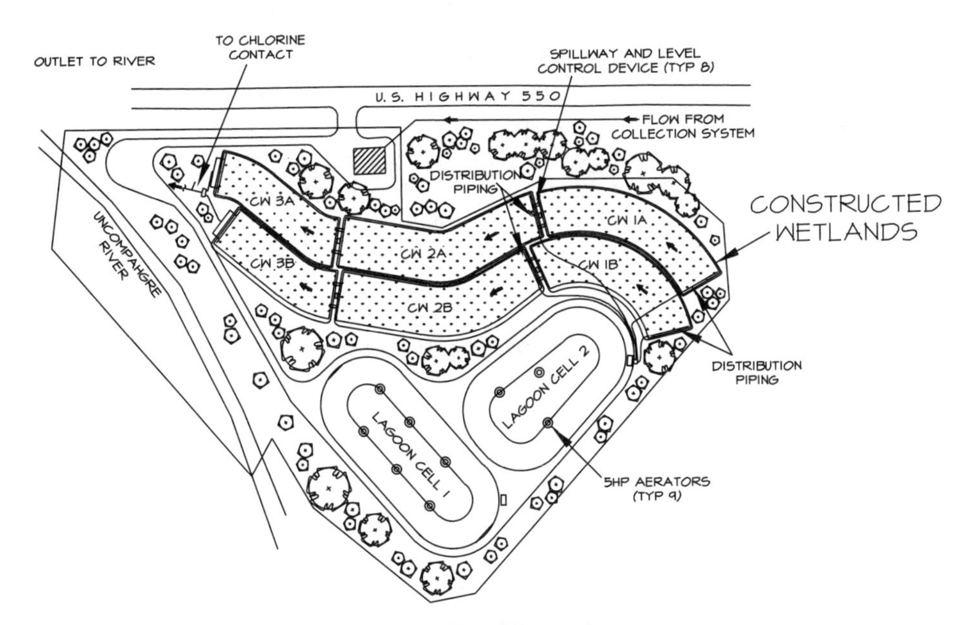

Figure 3.4 *City of Ouray plan area.*

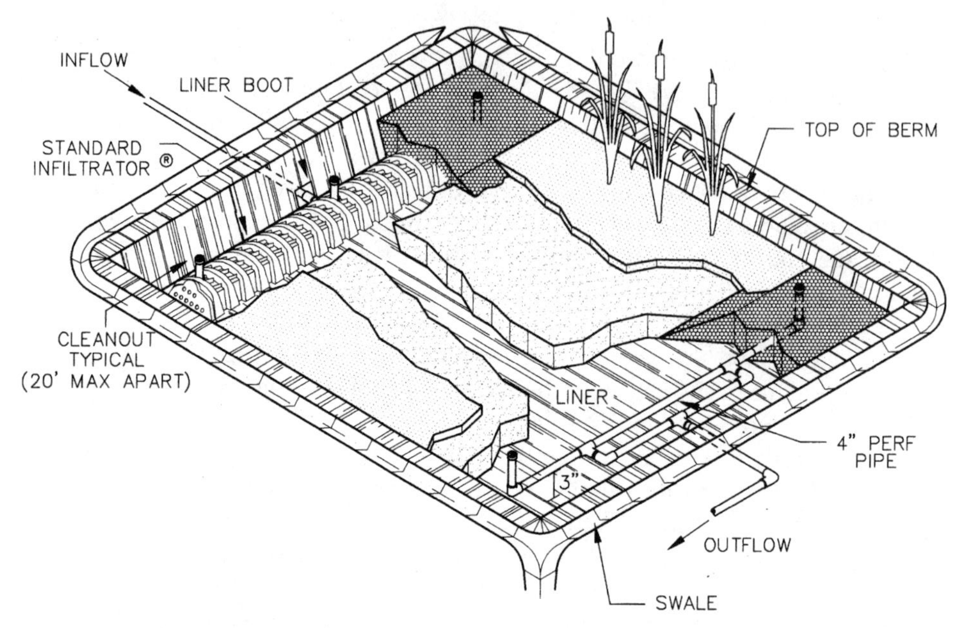

Figure 3.5 *Distribution piping (Courtesy of Southwest Wetlands Group).*

negligible. Equally important is that, when the pipe is underwater, freezing is highly unlikely.

Another design principle used here (Figure 3.5) is that there are multiple points of influent distribution. Instead of influent occurring only at the head of the wetlands, it can also be introduced into the second cell of each train. This minimizes the effects of high BOD loading by spreading the influent organic material over twice the area. Most of the solids will settle out within the first 30 feet from the distribution piping. By doubling these effective downstream areas, we can minimize the effects of the buildup of organic materials that may lead to anaerobic conditions in the wetlands.

Design Calculations for Ouray Constructed Wetlands Design calculations for Ouray's constructed wetlands are presented in Figure 3.6.

Performance As with all treatment system designs, it is important for the designer to remain in contact with the operator and the community to evaluate the design and ensure that there are no problems with the operation. Ouray's constructed wetlands wastewater treatment system was planned in November 1993, but no wastewater was introduced until March 1994. The data presented in Figure 3.7 represent the first year's performance.

1. Design Information (Complete information in shaded areas)

Location:	Ouray, CO				
Project Name:	City of Ouray As of:	12-Jun-05			
		Winter		Summer	
Flow	Q =	870.64	m3/day	1374.10	m3/day
	Q =	230000	gpd	363000	gpd
BOD in**	Co =	60	mg/L	60	mg/L
BOD out	Ce =	30	mg/L	30	mg/L
Nitrogen In	N =	5	mg/L	5	mg/L
Ammonia	NH3 =	1	mg/l	1	mg/l
Suspended Solids	TSS =	30	mg/l	30	mg/l
Water Temp In	Twi =	6	deg. C	15	deg. C
Water Temp Out	Two =	1	dg. C	14	deg. C
Ambient Air	Ta =	–25	deg. C	21	deg. C
Solar Radiation	SR =	550	btu/sf/d	2020	btu/sf/d

Other Information			
Bed depth*	d =	0.45	m
Porosity+	n =	0.7	(range 0.65 – 0.75)
Reaction Const	K20 =	0.678	/day

Notes:
*depth may be .45m or .3m
+porosity based on maximum density of plants

2. Calculations

A. Reaction Consant and Suface Area
(formulas 6.34 and 6.35, Reed, S.C., et al., "Natural Systems for Waste Management and Treatment")

		Winter		Summer	
Reaction Const.	Kt =	0.22408782		0.47796325	
Suface Area	As =	8549	sq. m	6326	sq. meters
	As =	92248	sq. ft	68259	sq. ft.
	As =	2.12	acres	68259	sq. ft.

B. Hydraulic Loading Rate (HLR) & Hydraulic Residence Time (HRT)
(formulas 9.11 and 9.2, ditto)

HLR =	10	cm/d	
HRT =	3.1	days =	74.0 hours

C. Nitrogen and Ammonia Concentrations in Effluent
(formulas 9.9 & 9.10 from WPCF Manual of Practice, FD-16)

TN =	5.00	mg/l
NH3 =	1.51	mg/l

D. Suspended Solids Concentrations in Effluent
(formulas 9.9 & 9.10 from WPCF Manual of Practice, FD-16)

TSS = 4.1 mg/l

Figure 3.6 *Design calculations.*

E. Calculate Dimensions & Velocity of Flow

Length to Width Ratio =		10	:1
	Width =	96	ft
	Length	960	ft
	Velocity	12.98	ft/hr

3. Calculate Winter Water Temperature
   (1st step: Calculate water temp at end of CW)

Distance (at end)	x =	960	ft	
Depth of Water	y =	1.48	ft	
water to air	Us =	0.264	btu/ft2/hr-degF	
Inf. Water Temp	Tw =	42.8	deg. F	
Ambient Air	Ta =	–13.0	deg. F	
Density of H20	delta =	62.4	deg. F	(Ground temp under wetlands)

		Winter	
Eff. Temp	Tw =	32	deg. F
	Tw = *	0.2	deg. C (must be > 3 deg. F)
Avg. Water Temp =		37.6	deg. F
(for N removal)		3	deg. C

Note: Insert Tw = * into water temp under design (line 12) values in an iterative process until Tw = * approx. equals influent water temp.

2nd Step: Water Temperature < 3 deg. C

Prop. Coefficient	0.024	ft/deg^.5*d^.5) for dense areas of vegetation
Time	3	days

Rate of Ice formation	0.16099689	ft/day
Total Thickness of Ice	0.2788548	ft

4. Calculate Water Loss/Gain in Constructed Wetlands
   A. Design Information (from NOAA)

Annual Pan Evaporation Rate =	35	in/yr
Average Annual Rain Fall =	39.31	in/yr

(Use greatest annual rainfall for past 20 years if available for monthly data)

	+ Inflow (gpd)	– Pan Evap. (in.)	+ Precip. (in.)	= Outflow (gpd)
Jan.	230000	0.39	2	233138
Feb.	230000	0.53	1.66	232546
Mar	230000	1.05	2.66	233376
Apr	230000	2.10	4.53	235463
May	363000	4.20	3.69	363612
June	363000	6.69	4.2	360800
July	363000	7.18	4.27	360182
Aug.	363000	6.20	3.75	360763
Sept.	363000	3.75	3.76	364464
Oct.	363000	2.03	3.51	366498
Nov.	230000	0.53	2.68	234332
Dec.	230000	0.39	2.6	234251
	Total	35	39.31	

Figure 3.6 *(Continued)*

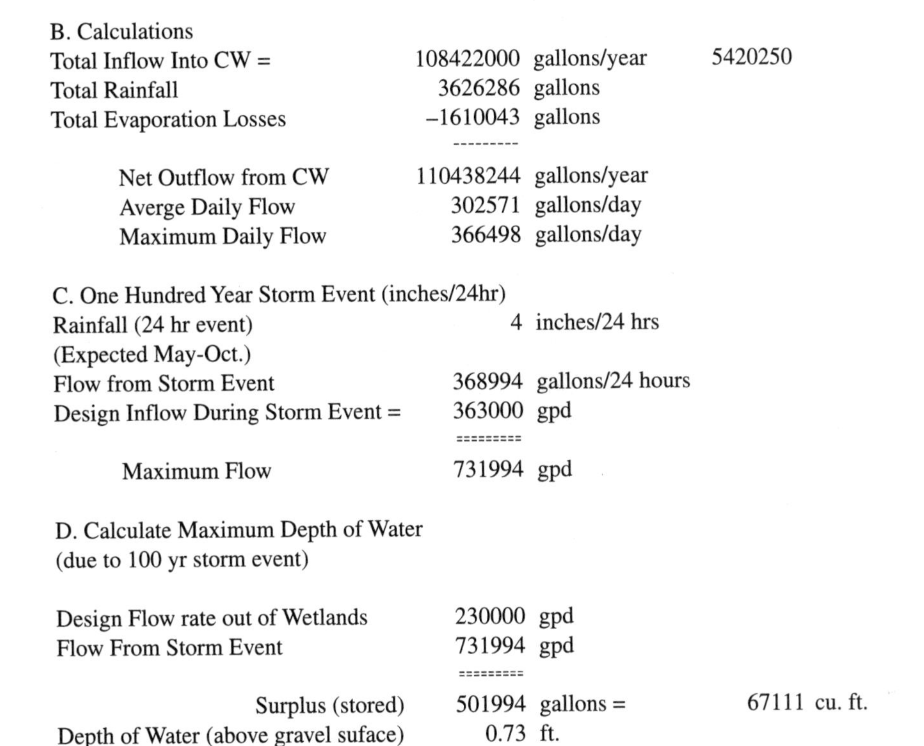

B. Calculations			
Total Inflow Into CW =	108422000	gallons/year	5420250
Total Rainfall	3626286	gallons	
Total Evaporation Losses	–1610043	gallons	

Net Outflow from CW	110438244	gallons/year	
Averge Daily Flow	302571	gallons/day	
Maximum Daily Flow	366498	gallons/day	

C. One Hundred Year Storm Event (inches/24hr)		
Rainfall (24 hr event)	4	inches/24 hrs
(Expected May-Oct.)		
Flow from Storm Event	368994	gallons/24 hours
Design Inflow During Storm Event =	363000	gpd
	=========	
Maximum Flow	731994	gpd

D. Calculate Maximum Depth of Water
(due to 100 yr storm event)

Design Flow rate out of Wetlands	230000	gpd	
Flow From Storm Event	731994	gpd	
	=========		
Surplus (stored)	501994	gallons =	67111 cu. ft.
Depth of Water (above gravel suface)	0.73	ft.	

Figure 3.6 *(Continued)*

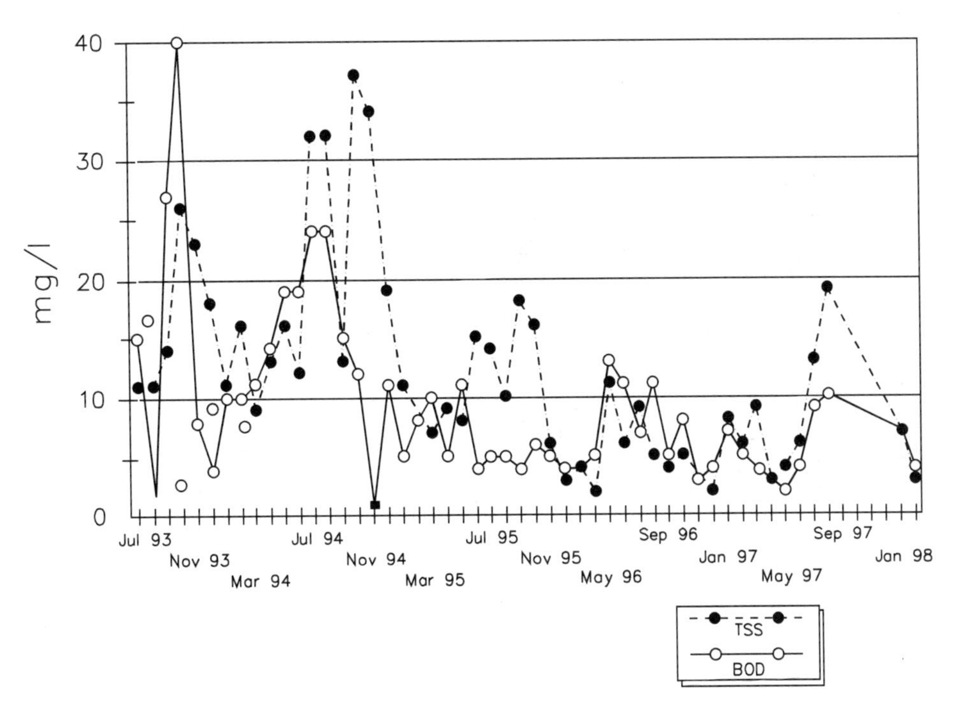

Figure 3.7 *Ouray WQ data.*

Village of Capitan, NM

Village of Capitan, New Mexico

PROJECT DESCRIPTION - Capitan is a small community located in the Sacramento Mountains. The existing sewage lagoons were filled with sludge and overflowing. Capitan needed a new treatment system capable of handling the existing wastewater flow, plus accommodate the expected population growth.

1. Design flow = 85,000 gpd
2. Zero discharge with land application option
3. Winter temperatures of -15 °F, summer temp of 95 °F
4. Provide habitat with U.S. Fish & Wildlife.

LIMITED FUNDS -The Village had limited bonding ability and required a low capital and operating cost system. The capital cost was $304,000 for 85,000 gpd system.

TECHNOLOGY- The system followed similar design concepts in Colorado and Arizona. The mayor and Village council had visited other small communities with similar systems and placed an RFP for firms with appropriate experience and qualifications.

DESIGN OBJECTIVES - Develop a cost effective system that is capable of treating 85,000 gpd and providing an aesthetic site development in the form of a lake and marsh.

- Meet the New Mexico water quality standards for land applied effluent and/or effluent discharge to ground water.
- Minimize the organic and nitrogen loading on the soil; reduce BOD to 20 mg/L; reduce nitrogen from 45 to less than 10 mg/L (average), as measured in monitoring wells.
- Incorporate ornamental vegetation to enhance the site and create an aesthetically pleasing project with wildfowl habitat and public park.

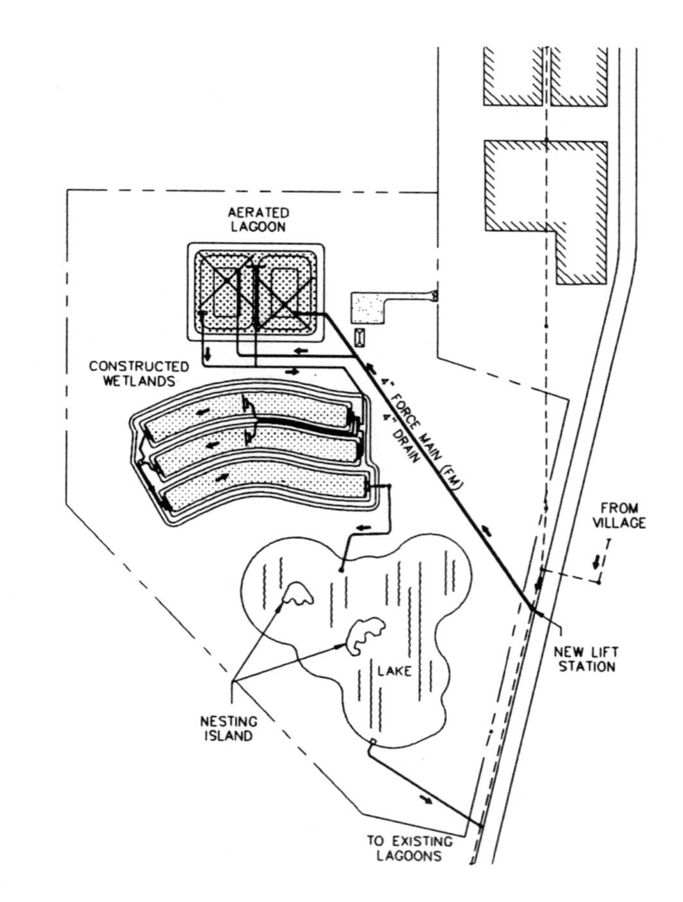

Hanson's Lakes, NE

Hanson Lakes, Bellevue, Nebraska

PROJECT DESCRIPTION - Approximately 40 years ago, an abandoned sandpit and lake property was subdivided into 300 individual lots. Landowners have incrementally constructed homes and made other improvements to create a permanent community. Typically, homeowners have located leach fields and water wells close together since their lots are quite small. This is threatening the quality of both groundwater and human health. Existing problems are:

1. The groundwater is relatively high (4 feet below ground surface) and it travels freely through the sandy soil. Water wells are quite shallow and intercept free moving ground water.
2. While tile fields drain well, they discharge relatively raw sewer effluent into the ground water. Free moving groundwater widely disperses pollutants throughout the water table.
3. People are drinking tainted well water and could easily become sick.
4. Appropriate State officials were not aware of Constructed Wetlands, and their benefits since none have ever been built in the State of Nebraska.
5. Limitations on Federal funding for constructing treatment plants makes Constructed Wetlands a viable and extremely competitive alternative to other solutions.

SOLUTION - In 1994, the community actively pursued the construction of a conventional treatment plant. Exposure to the potential value of Constructed Wetlands prompted the evaluation of 4 different solutions within a Feasibility Study. Options include:

- Constructed Wetlands
- Treatment by the City of Omaha ($5.4 million cost)
- On-site Mechanical Treatment Plant ($2.5 million cost)
- Oxidation Lagoon (Significant odor and land consumption problems, also lower efficiency)

The Constructed Wetlands option has been chosen due to Lifetime project costs of only $850,000 for the 20 year project life. Also the O&M costs are only $15,000 per year, with limited land consumption of 1.5 acres for 300 lots; creation of wildlife habitat and aesthetic benefits.

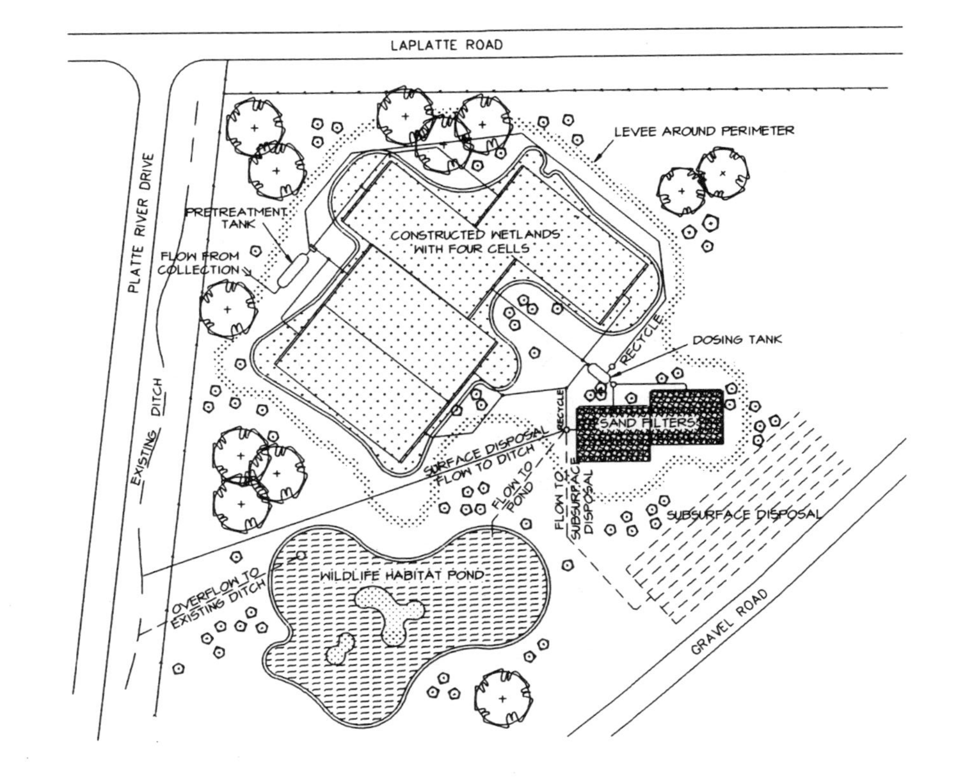

Town of Quilehtla, Tlaxcala, Mexico

PROJECT DESCRIPTION - The existing town of approximately 4000 people in the Municipal District of Zacatelco had a sewer system, but no wastwater treatment system. Quilehtla, located approximately 1 1/2 hours east of Mexico City, has both commercial, residential and agricultural wastewater characterized by a Biological Oxygen Demand (BOD) 4-8 times US standards for municipal wastewater. The wastewater was discharged into a seasonal watercourse without any treatment. The Governor's office needed a low cost system that would be easy to maintain and operate. A summary of the design problems include:

1. High strength wastewater, BOD =900-2000 mg/l; flow =85,000 gpd.
2. Limited capital budget
3. Simple to operate and maintain.
4. Minimize operational costs and use of electricity.
5. Provide on-site treatment of sludge.
5. Odor free; no mosquito breeding areas.

LIMITED FUNDS - Typical of most projects in Mexico, Quilehtla had a limited capital budget. Imports of equipment and materials would drain money from the State economy. If a wastewater treatment system could be built using local labor and materials, this would keep the money in the State and directly benefit the local economy. Equally important were the long term operating costs. Money paid to replace equipment or pay electricity costs is money lost to pay for local programs such as schools and hospitals.

DESIGN OBJECTIVES - Develop a system that blends with aesthetic site conditions while meeting engineering challenges caused by tight soils, hilly terrain, and limited space. Enhance site features and maximize opportunities to lower all costs to the fullest extent possible.

- Meet or exceed the Mexican WQ standards. (Actually meets US secondary standards.)
- Use local materials and labor. Minimize equipment. Minimize electrical requirements. (In fact the treatment process does not use any electricity. A lift station is required for bringing sewage to the treatment site.)
- Make an attractive odor free treatment facility that can easily be maintained by the local labor resources of a small Town in Mexico.

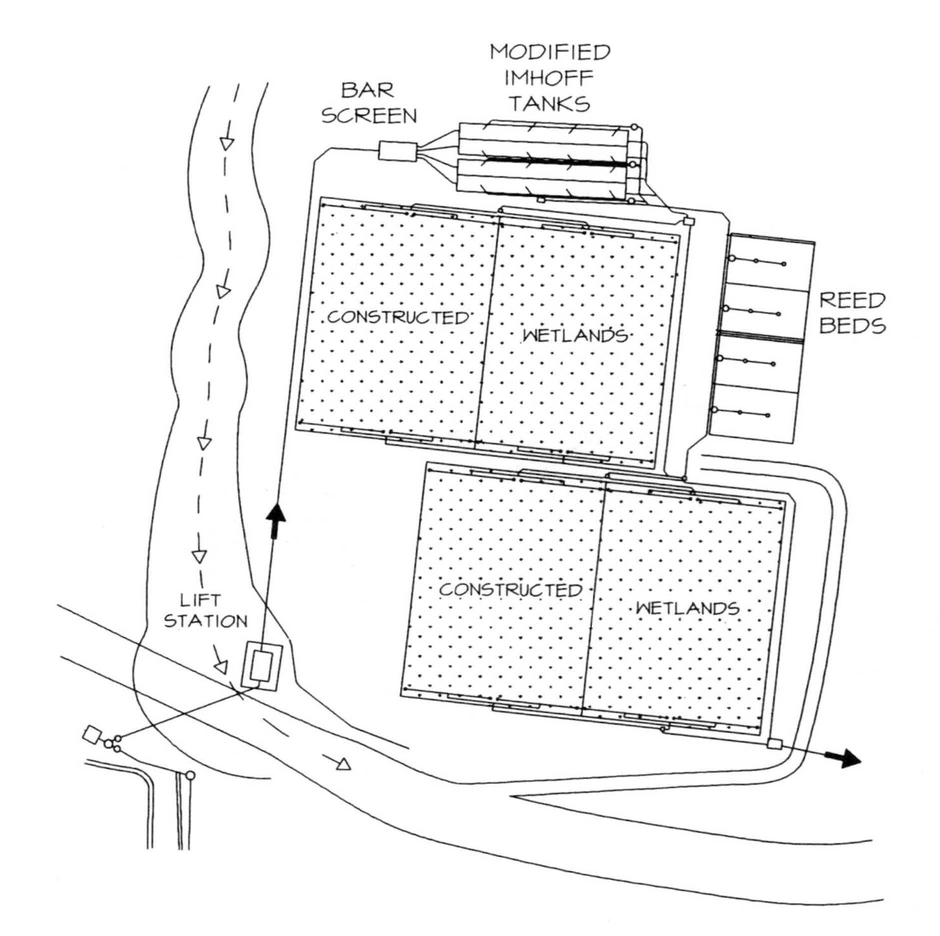

Firethorn Golf Course: Lincoln, Nebraska

PROJECT DESCRIPTION - The existing golf course and exclusive housing development treated its sewage with an on-site leach field using a perforated plastic pipe. The system did not operate correctly for several years due to high watertable and tight clay soils. Untreated effluent erupted through the ground. Connecting the sewer into the existing municipal system proved to be cost prohibitive, as did the O&M costs for an on-site mechanical system. A summary of the problems include:

1. Failing system - caused by a high watertable and clay soils.
2. Remote site - long distance between houses and clubhouse.
3. Impractical to tie into the city sewer main - extremely high cost.
4. Expensive on-site mechanical system - O&M costs too high
5. Concerns about property value - select a system that enhances the site.

LIMITED FUNDS - Firethorn Golf Course wanted to build a low cost system to minimize the financial impact of a second sewer system upon residents. The primary benefit will be the extremely low O&M costs. Also, the State of Nebraska maintains an interest in Constructed Wetlands. The combination of lower construction and O&M costs makes Constructed Wetlands an extremely attractive treatment alternative to the State.

LANDMARK OPPORTUNITY - The Firethorn project is the very first Constructed Wetlands treatment system to be built in Nebraska. The public will benefit when they are able to treat their wastewater at extremely low costs and conduct system maintenance locally.

DESIGN OBJECTIVES - Develop a system that blends with aesthetic site conditions while meeting engineering challenges caused by tight soils, hilly terrain, and limited space. Enhance site features and maximize opportunities to lower all costs to the fullest extent possible.

- Meet the Nebraska state water quality standards for effluent water being discharged into the creek. (Avg CBOD = 5 mg/l, avg TSS = 5.5 mg/l)
- Minimize capital, energy and maintenance costs.
- Use local materials.
- Incorporate prairie grasses and other ornamental vegetation to blend with the golf course and create an aesthetically pleasing project.

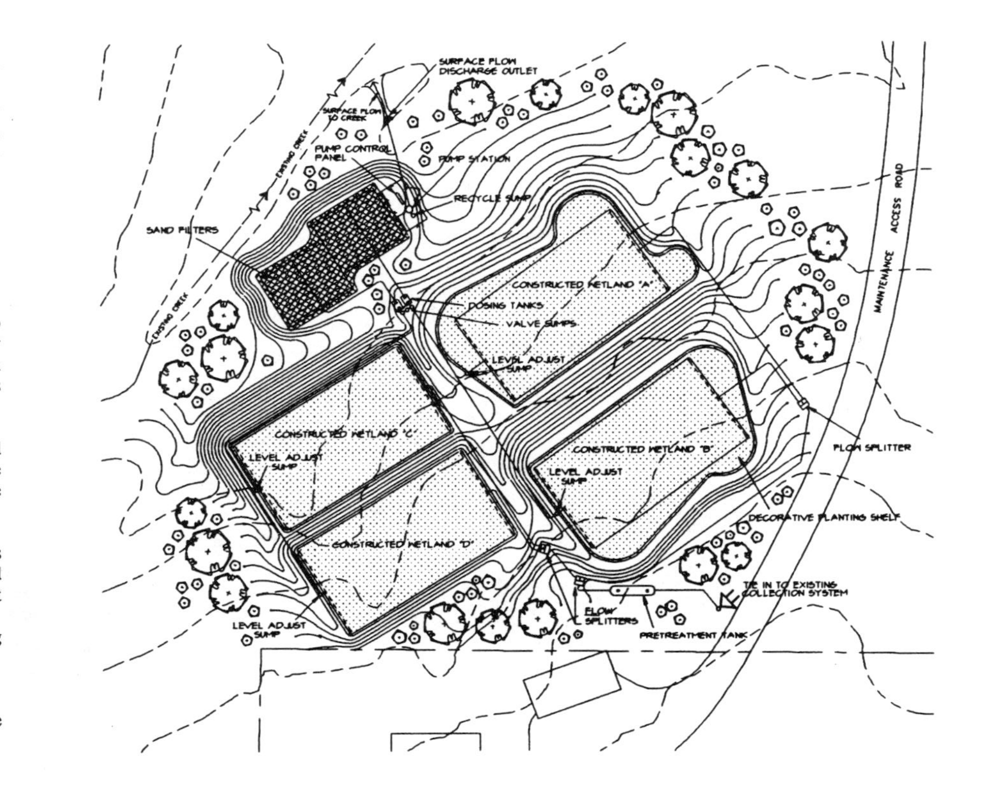

Ten Stones Subdivision, Burlington, VT

PROJECT DESCRIPTION - This is a new subdivision located south of Burlington. The original design called for a dosed field with a large area set aside for replacing the eventually failing field. As an alternative, a constructed wetlands was proposed to treat the effluent from each of 13 septic tanks (design flow = 3900 gpd), and then dose the field.

1. Shallow soil cover above bedrock.
2. No community sewer system nearby. Impractical to tie into the city sewer main - extremely high cost.
3. Traditional dosed field requied set-aside.
4. Concerns about property value - select a system that enhances the site, and is environmentally friendly.

LIMITED FUNDS - The developer wanted to build a low cost system to minimize the financial impact of on-site treatment. The primary economic benefits to the homeowners will be the low O&M costs.

INNOVATIVE TECHNOLOGY- The Ten Stones project is the very first Constructed Wetlands treatment system for a subdivision to be built in Vermont. Data will be collected to allow regulators to monitor the system and understand the effects on soil loading.

DESIGN OBJECTIVES - Develop a system that blends with aesthetic site conditions while meeting engineering challenges caused by tight soils, hilly terrain, and limited space. Enhance site features and maximize opportunities to lower all costs to the fullest extent possible.

- Meet the Vermont state water quality standards for effluent water being discharged into the ground.
- Minimize the organic and nitrogen loading on the soil; reduce BOD from 160 to 10 mg/L, and nitrogen from 45 to 20 mg/L (average).
- Incorporate ornamental vegetation to enhance the site and create an aesthetically pleasing project.

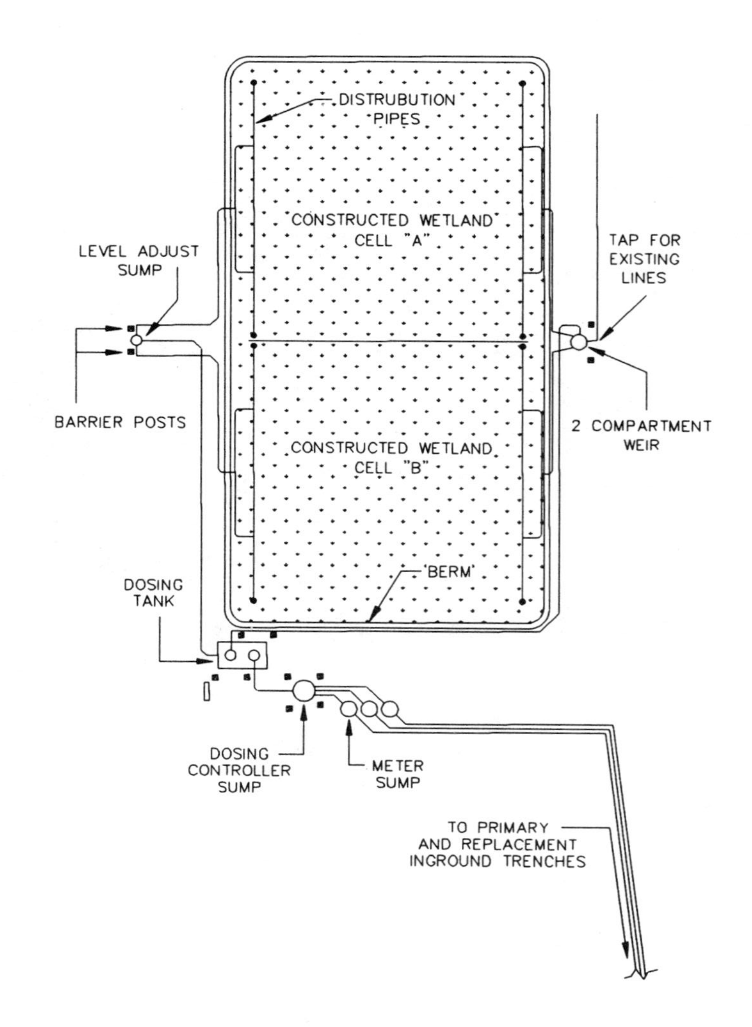

DESIGNING FOR EXISTING FACILITIES

Designing for existing facilities usually involves bringing existing wastewater treatment facilities into compliance by providing additional treatment or by designing completely new facilities to replace the old ones. The design process is essentially the same as for new facilities, except that the existing wastewater flows can be measured and characterized. Estimating the ultimate design flow is still a major challenge and is usually the first task. Existing facilities will often need to be upgraded because the hydraulic capacity of the original design has been exceeded or because more stringent water quality requirements have been imposed by the regulatory agency.

In either case, constructed wetlands are often a good choice for small communities because they allow existing facilities to be used in conjunction with wetlands. Often, for example, small communities have wastewater stabilization lagoons or aerated lagoons that no longer meet the required water quality standards because of population growth or because water quality standards have become more stringent. Too much wastewater may be flowing through the lagoons, and consequently the BOD and TSS will exceed the permit limits. The addition of the wetlands will allow the existing lagoon to be used for primary treatment with the wetlands used for secondary and tertiary treatment.

The entire process of designing new facilities or replacing existing facilities has been described in great length and in excellent detail in the U.S. EPA publication *Small Community Wastewater Treatment and Collection Systems* (U.S. EPA, 1992). Besides providing an excellent description of the costs, advantages, and disadvantages of various options, the publication describes in detail both the designer's and the community's responsibilities. Although much of the planning involves common sense, the technical descriptions are excellent and the design issues are clearly identified. This publication should be read prior to undertaking any design.

The design questions that should be answered after determining the wastewater flow are the following:

1. Can any parts of the existing facility be incorporated into the new facility? Should we repair or replace the existing facility?
2. Is the site large enough for the proposed expansion? If not, are other sites available? Can we pump the wastewater to some other nearby existing treatment facility?
3. What potential wastewater treatment alternatives would work?
4. What are the capital and operating costs for each alternative?
5. How will the facility be paid for by the community?

Assuming that constructed wetlands are the preferred alternative, site considerations become the next important issue. The U.S. EPA publication *Constructed Wetlands for Wastewater Treatment and Wildlife Habitat* (U.S. EPA, 1993a), provides some excellent examples of small-community systems that have provided not only improved water quality but also open space, wildlife habitats, and community parks.

The design of wildlife habitats, public parks, and wastewater treatment facilities requires that the engineer, biologist, and landscape architect work closely together throughout the project to ensure that the requirements for each element are properly addressed and that any compromises made do not jeopardize the ultimate success of the project. The premise for the design of such community systems is that the existing wastewater flow is in fact a resource, if properly understood, and that any upgrade to an existing facility must acknowledge the potential for habitat development. In fact, in most areas of the West, wastewater is the only water available for habitat development. Therefore, in the cost/benefit analysis, some value must be placed on open space, habitat, and wildlife.

If available land is an issue, then habitat and park space may have to be sacrificed; however, these elements should not be discarded easily because of land availability, economics, or the inability of the designer to understand all of the issues. It is essential to get help. Wetlands for habitat require a multidisciplinary approach that can be effectively implemented only with the assistance of the appropriate experts.

Site aesthetics are not normally part of wastewater treatment facilities; however, constructed wetlands and wetlands habitat provide a rare opportunity for the landscape designer and wildlife biologist to design a wastewater treatment facility that is an attractant rather than a nuisance. Again, the examples in *Constructed Wetlands for Wastewater Treatment and Wildlife Habitat* should provide some excellent design ideas for the replacement wastewater treatment facility.

DESIGN FOR INDUSTRIAL, MINING, AND AGRICULTURAL SYSTEMS

Wastewater treatment for industrial, mining, and agricultural systems has been accomplished using wetlands. The problems presented by these types of wastewater are potentially much greater than those involving residential and small-community wastewater. However, the design process is essential the same as that described in the section "The Design Process for New Facilities," and in essence, the task is to characterize the wastewater flow in terms of both volume and constituents.

In general, the process of characterization is more complicated because of the particular problems presented by, for example, pH values that are very low or very high, metals, nutrient deficiencies or excesses, biologically toxic compounds, highly odoriferous compounds, high solids content, and large seasonal fluctuations. Characterization becomes a specialized effort after requiring the advice of a toxicologist, as well as a sophisticated laboratory. Because of potential nutrient deficiencies, the assistance of a microbiologist may also be warranted.

In spite of the complications presented by industrial, mining, and agricultural wastewater, the designer should understand that almost all of the industrial pollutants can be biologically remediated by bacteria, plants, and fungi, and that the best environment for creating the ideal conditions for the remediation process is wetlands. Many of our ore deposits are microbial deposits that occurred in ancient wetlands. The same microbial processes that created the Mesabi iron ore or the copper deposits of New Mexico and Arizona can be put to work in treating acid mine drainage to remove metals and neutralize the acid.

Some of the plants used for the treatment of acid mine drainage will be different from those used in municipal wastewater wetlands. Plants that are more tolerant of acidity, such as bog wetlands plants, are more successful in the initial stages of treatment. Changing the pH is the primary challenge; this is accomplished by adding carbon in the form of dead vegetation and creating open-water wetlands where algae can thrive. The daily respiratory cycle for algae will dramatically change the pH. As the pH rises, metals are precipitated out by bacteria, as sulfides and oxides; except at very low pH values, these metallic compounds are generally no longer soluble.

Understanding these principles requires a great deal of complex knowledge that is usually not available to one individual. Cooperation with members of other disciplines is essential. Some general rules for design are as follows:

1. Create the biological conditions that allow the pH to move toward the mean.
2. Ensure that all of the major nutrients are present in the appropriate ratios. If nitrogen is too high, create the conditions for its removal or recycle it. If the nitrogen concentration is too low, add nitrogen in the form of ammonia. The same applies to phosphorus.
3. Always add micronutrients.
4. Seed the wetlands with bacteria drawn from the edges of sites contaminated with the pollutants to be removed. Genetic drift and adaptation will always provide a set of bacteria that are already doing the job.
5. Select plants appropriate to the region and the desired water quality. Usually, someone has polluted a body of water with the same materials you will be dealing with. Select those plants. Whenever possible, duplicate nature.
6. Get assistance from the following experts:
   a. Wetlands ecologist.
   b. Microbial ecologist.
   c. Plant toxicologist.
   d. Phycologist (a specialist in the study of algae)

Figure 3.8 is a plan view of a treatment facility for industrial wastewater from an egg processing facility. Note the provisions for the addition of micronutrients and recycling.

DESIGNING FOR STORMWATER RUNOFF AND COMBINED SEWER OVERFLOWS

Stormwater runoff from municipal streets and combined sewer overflows are characterized by intermittent, nonuniform flows, usually with a great deal of seasonal variability. Summer thunderstorms produce a very different pattern of runoff than snow storms.

Stormwater runoff is also characterized by variable water quality parameters. The major pollutants of concern are metals and petroleum hydrocarbons. Organic material, suspended solids, and human fecal material (in combined sewer outflows) generally occur in ranges that are less than 50 percent of those of municipal sewage. Water quality improves as the storm progresses. The first flush from the streets at the beginning of the rainy season has the highest levels of volatile organic com-

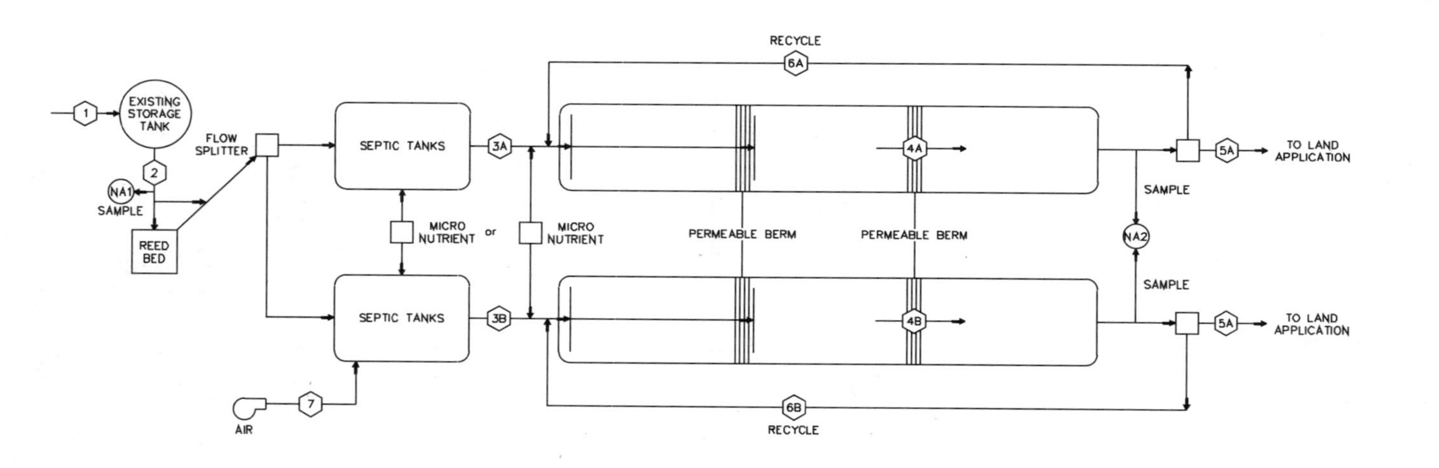

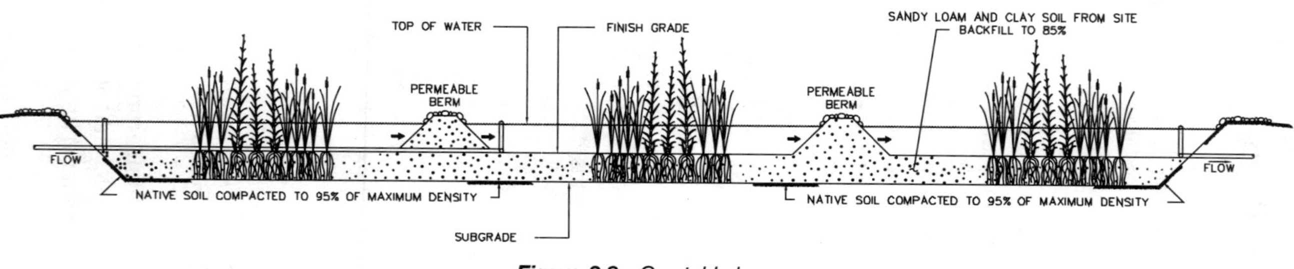

Figure 3.8 *Crystal Lake.*

pounds and metals. As the rainy season progresses, pollutant concentrations decline.

Because the areas served by storm drains in most cities are very large, it is not practical to attempt to treat all of the wastewater. A 1-inch rainfall on a square mile of a city will produce runoff in excess of 17 million gallons. It is neither possible nor necessary to treat this entire volume. Clearly it is much more sensible and economic to treat the initial runoff.

How do we determine the amount of runoff that does need to be treated? The first step is to sample runoff from various storm events. Every location in the United States has standard storm events that can be used to determine the volume of a particular storm. However, the water quality will be different for each event. The worst-case events are those that require sampling; these events are the first rains of the rainy season or the runoff after the first snowstorm. As an alternative, simulation of storm events can be performed using desktop computers.

Each type of event should be sampled for the following:

Volatile organic compounds
Metals
TSS
BOD
Fecal coliform bacteria
pH
Alkalinity

Potential sites need to be investigated. At a minimum, the following factors should be studied:

Potential for flooding
Access to the site for existing storm sewers
Soils, seasonal high groundwater
Land area available

Once a potential site is selected, we can consider what to accomplish. The easiest objective is the removal of TSS and organic material. This can be accomplished by uniformly distributing some initial portion of the storm event over the wetlands. The wetlands must be designed to avoid short circuiting, and overloading. For example, if the system is designed for a 2-year/24-hour storm event and then a 10 year event occurs, provisions must be made to divert the excess. As a design goal, BOD and TSS standards can be met by retaining the storm water for two days.

Removal of volatile organic compounds, in particular petroleum hydrocarbons, is very good in surface flow wetlands. The major removal mechanism is volatilization, although there is also good biological treatment. Volatilization and biodegradation are temperature-dependent processes, occurring more rapidly in the summer than in the winter. Petroleum hydrocarbons are attracted to the stems and leaves of the wetlands plants, as well as to the dead vegetation and the underlying soil. As the vegetation increases and the leaf litter builds up, efficiency improves.

Essentially, the wetlands are a biological filter utilizing both biological and mechanical processes. Metals can also be removed in constructed wetlands. However, some distinction must be made concerning the form of the metals. Most of the metals washed off of city streets will be in the metallic form which settle out as solids in constructed wetlands. However, some metals become solubilized, especially in the presence of acid rains. These ionic metals are more difficult to remove, but they can be removed by microbial activity in the wetlands, which causes them to precipitate out of solution as metallic sulfides or oxides.

One important design factor in stormwater runoff is making adequate provisions for sediments and trash. The treatment facility should have a trash rack and a sedimentation basin that are accessible to equipment for periodic cleaning. The trash racks should be designed with an overflow bypass in the event that they become completely clogged with trash.

In designing wetlands, particular attention should be given to good distribution of the wastewater flow. More detention time is better. Equalization ponds can be included to allow the storm to be discharged over a longer time span. This will allow the wetlands to become smaller. As an example, suppose that the typical storm event is 20 million gallons in twenty-four hours. We have decided that we need to treat the first 1 million gallons. In order to minimize our land requirements, we decide to build a 1 million-gallon equalization basin that will discharge our storm over five days. The design flow is therefore 1,000,000/5 = 200,000 gpd. If we decide that our treatment time is two days (minimum), then our wetlands area is:

$$As = (2 \text{ days} * 200{,}000 \text{ gal/day}/7.48 \text{ gal/cf})/1.5 \text{ ft} = 35{,}650 \text{ ft}^2$$

Passive treatment strategies are clearly the most cost-effective, provided, of course, that land is available. Equally important, stormwater wetlands can be incorporated into riparian habitat restoration projects, river walks, and "Rails to Trails" projects. The vegetation used in stormwater wetlands provides an excellent habitat for wildlife.

THE ECONOMICS OF CONSTRUCTED WETLANDS

Subsurface Flow Systems

A variety of factors will affect the cost of wastewater treatment wetlands:

- Detention time (will be climate dependent)
- Treatment goals
- Depth of media (deeper systems require less liner)
- Type of pretreatment
- Number of cells (more cells require more hydraulic control structures and more liner)
- Source and availability of gravel media
- Terrain

Due to the wide range of design possibilities, some standardized basis for comparison must be established. The basis for comparison includes the following:

Construction cost versus detention time (HRT)
Construction cost versus gallons treated per day

Additional complications in providing good comparative cost information include separating different design goals, such as BOD removal versus BOD and nitrogen removal. Effluent BOD design goals will also affect detention times. For example, an effluent BOD of 30 mg/L versus one of 10 mg/L will increase the area by 100 percent.

Detention time is a good means of comparison because of the different depths designers have used. Design depths typically range from 12- to 30-inches. As an example, a system designed with a 12-inch depth will have twice the area of a system with a 24-inch depth.

Economic efficiencies can be measured in other terms, such as pounds of BOD (or kilograms) removed per day, or a comparison can be made between treatment options. Other comparisons that are useful are energy costs, as suggested originally in Table 8.6 of the EPA publication *Process Design Manual: Land Treatment of Municipal Wastewater*. (U.S. EPA 1980). A modified version is reproduced in Table 3.3. Whichever method is used, the following must be recognized: wetlands are a temperature-sensitive technology, and comparing system costs in Nebraska to those in Florida is like comparing the price of corn to the price of oranges. Ultimately, cost comparisons should be based on systems with similar discharge limits located in similar climate zones.

As Table 3.3 shows, there is an increasing energy cost as systems move from land intensive to mechanical systems. Water quality does not necessarily improve with increased energy until the use of activated sludge, and then the improvement

TABLE 3.3 Energy Costs of Various Wastewater Treatment Technologies to Treat 1 Million Gallons per Day (U.S. EPA, 1980)

	Effluent Quality			
Treatment System	BOD	SS	N	Energy 1000 kWh/yr
Rapid infiltration/facultative pond	5	1	10	150
Slow rate, ridge & furrow	1	1	3	181
Anaerobic digestion/wetlands/slow rate	1	1	3	185
Facultative pond/sand filter	15	15	10	241
Aerated pond/sand filter	15	15	20	506
Extended aeration + sludge drying	20	20	—	683
Extended aeration + sand filter	15	15		708
Trickling filter + anaerobic digestion	30	30		783
RBC + anaerobic digestion	30	30		805
Trickling filter w/nitrogen removal	20	10	5	838
Activated sludge w/anaerobic digestion	20	20		889
Activated sludge + nitrification	15	10		1,051
Activated sludge w/incineration	20	20		1,440
Activated sludge w advanced treatment	<10	5	<1	3,809
Physical/chemical advanced secondary	10	10	1	4,464

is in nitrogen. In order to minimize the land requirements, energy is required. To go from a 28-acre wetlands and land application system treating 1 million gallons per day to an activated sludge system occupying 1 acre, requires 3,809 − 185 = 3,624 × 10^3 kilowatt hours per year. At 7 cents per kilowatt hour, this is an annual cost difference of $253,680. Over a twenty-year life, the total difference is $5.1 million. The energy cost of the land is $187,900 per acre. This is expensive land.

When reviewing costs, the engineer must consider other factors, such as the number of cells and the scale. For example, four cells totaling 1 acre will cost more to build than one cell totaling 1 acre; large systems will cost less per gallon than small systems.

Pretreatment will have a major effect on wetlands construction costs. If the collection system is a small-diameter sewer system with interceptor tanks, pretreatment will produce an influent BOD of about 120 to 140 mg/L with anaerobic properties (see Table 4-1, of the U.S. EPA publication *On-site Wastewater Treatment Systems.* Influent TSS averages 30 mg/L in some of these small-diameter systems. Compare this to a partial-mix aerated lagoon where the influent into the constructed wetlands is likely to have BOD in the range of 30–60 mg/L and TSS in the range of 60 to 80 mg/L. The point is that a wetlands receiving BOD of 50 mg/L will not be the same size as a wetlands receiving 130 mg/L, given equal flows of wastewater.

The first system will require additional treatment steps if nitrogen removal is part of the design goal. Anaerobic pretreatment will produce nitrogen principally as ammonia, which will not nitrify very efficiently in subsurface wetlands, especially in cold weather. However, aerobic pretreatment (aerated lagoon) will produce nitrogen principally as nitrates, which will easily be denitrified in wetlands. The total energy cost of the second system is considerably higher than that of the first, but the capital cost of the wetlands is significantly less. Obviously, both total system costs and treatment goals must be considered.

Capital Costs

Wetlands costs can be broken down into the following components:

Excavation
Liner
Gravel
Plants
Distribution and control structures
Fencing
Other

Given equal detention times, three components clearly encourage the designer to use deeper rather than shallow wetlands. As the wetlands get deeper, the liner area declines, fewer plants are required, and perimeter fencing becomes smaller. Gravel and excavation costs remain the same. Design depth will also affect heat losses and wintertime performance. Is treatment affected by deeper design depths? Should the designer always opt for those plants that have the greatest and deepest root penetration? Cattails have the least root penetration but are the easiest plants to propagate and consequently the least expensive plants. Conversely, reeds have the deepest root penetration but are very difficult to transplant into gravel beds and are therefore the most expensive plants.

The single most important factor affecting the capital cost is the cost of gravel, followed by the cost of the liner material. The costs of both items will increase as the specifications become more severe, for example, increased liner thickness, decreasing the percentage of fines (i.e., material passing the 200 sieve) in the gravel.

As a rule, gravel is 40–50 percent of the cost of a system for a 50,000-square-foot system, with the percentage increasing as the system is enlarged. The reason for this increase is that other costs decrease as the system grows. For example, the area of the perimeter runout material in the liner decreases as a percentage of the total area. Perimeter fencing costs decline for the same reason.

Gravel usually costs about $9.50 per ton or $18 per cubic yard throughout the United States. Hauling costs can add significantly to the project, and delivered costs can exceed $20 per cubic yard. There are also many areas in the United States where gravel is simply not available or is very costly. Some states, such as Florida, are considering the use of recycled concrete rubble.

Liners generally run 15–25 percent (see Table 3.4) of the total cost, with this percentage declining as the system gets larger. The percentage rises as more expensive liner materials are used and the number of cells increases. Soils with large angular rocks ($>1^{1/2}$ inches) may require the use of an underlayment such as geotextiles or sand. If design requirements dictate the need for geotextiles, costs will increase by 5 to 8 cents per square foot. If river run gravel is not available, sand or geotextiles should be placed on top of the liner.

Liner costs are based on the quantity, thickness, and type of material specified. A good argument can be made for eliminating liners in certain soils with a high clay content, but as regulators focus more attention on groundwater, reliance on use of in situ soils becomes problematic. Even with good soils in place, the cost of testing and compaction can exceed the cost of a 30-mil polyvinyl chloride liner.

Liner costs have decreased significantly in recent years because of demand and because most liners are petroleum-based products. These current low prices could easily rise, as they did during the oil price increase of the 1970s, which would make clay soils and bentonite much more competitive. It is possible to install a 30 mil polyvinyl chloride liner for 38–40 cents per square foot in small systems, and for 30–35 cents in systems with 100,000 square feet or more. These price ranges are generally available throughout the United States. Experienced liner crews will travel almost anywhere the job requires.

Table 3.4 provides a good estimate of the cost of installing liners for quantities greater than 100,000 square feet. These costs are for labor and the materials specified. Prices will be higher in the Northeast and California. These prices are based

TABLE 3.4 Cost of Liners ($/sf, 100,000 ft² min.)

PVC 30 mil	PE 40 mil	PPE 40	Hypalon 60 mil	XR-5	Reinforced PPE 45 mil
0.30–0.35	0.35–0.40	0.45–0.50	0.60–0.70	0.90	0.55

on competitive bids by liner installers who have installed more than 2 million square feet.

The use of clay and clay with scrims or the use of in situ soils presents some problems that are usually site specific and make estimating the costs of limiting percolation difficult. The addition of bentonite to the soil or the use of in situ clay soils is at first glance very simple. However, the additional costs of testing and compaction can add significantly to the cost of the project. Often the net result is that the costs associated with bentonite or in situ soils are the same as the cost of installing a PVC liner. The engineer must carefully evaluate testing costs, the additional time required, and equipment costs associated with mixing, wetting, and compaction.

Clay liners made of a thin layer of bentonite, or clay placed between two sheets of polyester fabric, or alternatively polypropylene geo-textiles, have been used in landfill applications and in at least one subsurface wetland. Clay liners with scrims have properties similar to those of XR-5 and Hypalon or reinforced polypropylene (PPE), except for weight. Shipping costs can make clay liners more expensive than PVC or PPE; however, installation is very simple, and the resistance to chemical degradation is equal to that of Hypalon or XR-5. If shipping distances are reasonable, prices including installation have been quoted at 38–40 cents per square foot.

Excavation/earthwork is generally the third or fourth largest cost of a typical project. Obviously, this cost is terrain dependent. Flat sites in Nebraska on sandy loam will be easier to excavate than sites on mountainsides in Colorado. Given this obvious caveat, most subsurface wetlands sites are usually constructed on level sites with good soils. As a result, excavation costs are usually about $1.50 to $2.50 per cubic yard.

Plants are generally a minor cost. The plants used in subsurface wetlands (cattails, reeds, and bulrushes) are usually available everywhere in the United States. Occasionally, they can be collected from local sources and planted in the constructed wetlands. In some cases, planting has been coordinated with county drainage ditch-cleaning operations, so that the cost for plants to the project is zero. Only planting labor is required. Planting in gravel can be accomplished easily with experienced crews planting 600 to 1,000 plants per person per day. However, if the project must bear the costs of harvesting, separation, cleaning, and transport to the job site, then the plants are likely to be very expensive. The alternative is to seek wetlands nurseries that are capable of providing the quantity, species, and quality of plants that the job requires.

Because of wetlands mitigation work, there are now many nurseries throughout the United States capable of growing and planting the plants used in constructed wetlands. The advantage of a nursery operation is that large quantities of viable plants 12–18 inches high can be grown for ease of harvest and subsequently transplanted by hand or machine, with a very high degree of transplant success. Designers can and should expect a minimum of 80 percent survivability. Costs will usually run about 50 cents to $1 per plant, with most bids at the lower end. Many nurseries will also grow plants on a contract basis for a particular project. By planning ahead, the designer can obtain discounts on plants.

The issue for the designer is plant spacing. Plants placed in 3-foot centers will each have to grow to fill 9 square feet while plants on 18-inch centers must fill only 2.25 square feet. For a 50,000 square-foot wetlands, at 50 cents each, plant costs are $2,800 and $11,000, respectively. The problem for the designer is that a 20 percent loss at 3-foot centers means that there will very likely be large unvegetated areas. These will eventually fill in, but can the project wait for the next growing season for these areas to fill in naturally or be replanted? The $8,000 difference on a project of this size may not be worth the wait.

In the past, planting was a casual affair, with success relying primarily on the hardiness of the plant species. Cattails and reeds, once started, are very aggressive and almost impossible to eliminate. Areas that were devoid of vegetation are not particularly important in large-scale subsurface projects, but unvegetated areas on small projects need to be replanted as soon as possible. Replanting is a definite cost consideration and can be included in the specifications requiring a certain minimum survivable population of plants. Experienced nursery personnel are capable of meeting these specifications and can be called upon to replant by contract if necessary. The designer should expect the same type of performance on this part of the contract as she or he would expect from pump suppliers or liner installers.

Other minor costs include piping costs and level control structures, flow distribution structures, flow meters, and fencing. In addition, reseeding and erosion control costs should be provided for in any design. Piping materials are generally plastics such as polyvinyl chloride (PVC), polyethylene (PE), and alkylbutylsulfate (ABS), commonly available throughout the United States. Plumbing costs will be between 6 to 7 percent. Level control and flow distribution structures can be built out of concrete block, cast-in-place and precast concrete; for smaller systems, reinforced PVC units are commercially available. Depending on the number of cells, these type of structures will usually run about 5 to 6 percent of the total cost.

Table 3.5 provides a good estimate of the relative costs for a 50,000-square-foot system, 2 feet deep, using typical unit prices that can be found in many places in the United States. Note that these costs are for the wetlands only. They do not include costs for pretreatment facilities or for the disposal of the treated effluent.

Table 3.5 shows typical costs. These should be considered similar to the concept of standard construction units presented in cost estimating handbooks. The caveats associated with this approach (e.g., site conditions, distance to the gravel pit, liner requirements, and water quality) require another approach to estimating the costs of construction of subsurface flow wetlands.

Since the time the U.S. EPA data base was published, many additional constructed wetlands have been designed and built, especially under 50,000 gpd. The costs for these systems can be presented in two ways.

TABLE 3.5 Cost of a Typical 50,000 Square-Foot Subsurface Constructed Wetland

Description	Units	Unit Price	Total $	Percent
Excavation/compaction	cubic yard	$1.75	13,000	10.7
Gravel	cubic yard	$16.00	51,900	42.6
Liner—30 mil PC	sf	$0.35	19,250	15.8
Plants—18″ o.c.	each	$0.60	13,330	10.9
Plumbing	ls		7,500	6.1
Control structures	ls		7,000	5.7
Other	ls		10,000	8.2
		Total	$121,980	

Figure 3.9 presents the costs of building wetlands as function of dollars per day of detention time, and Figure 3.10 and 3.11 present the information as a function of dollars per gallon per day. Costs are presented this way because designs have different treatment objectives (e.g., BOD removal or BOD and nitrogen removal) and because two similar flows with different design objectives will have very different costs.

Figure 3.9 is based on the assumption that as the system detention time increases, water quality increases. Once design objectives and hydraulic residence time are determined, costs can be estimated using this graph. A linear fit with the value of

$$\text{Construction cost (\$)} = .91 * X + 2{,}150$$

where X = flow (gpd) * detention time (days)

has an r^2. of .64. Although this is a poor correlation, it does give a rough estimate of costs. For example a 20,000 gallon per day system with three days of treatment will cost $56,750. Figures 3.10 and 3.11 show costs for constructed wetlands in the range of 0 to 1 millon gallons per day and 0 to 200,000 gallons per day. It is important to note that the larger systems were built without liners; it is difficult to get less than $2 per gallon per day with liners. Note, however, the logarithmic nature of the curves in Figures 3.10 and 3.11. As the hydraulic capacity of the system increases, the costs per gallon treated per day decline.

Operation and Maintenance Costs

Wetlands are almost invariably one part of a multipart treatment system. Determining actual operating costs from the database is therefore difficult because the

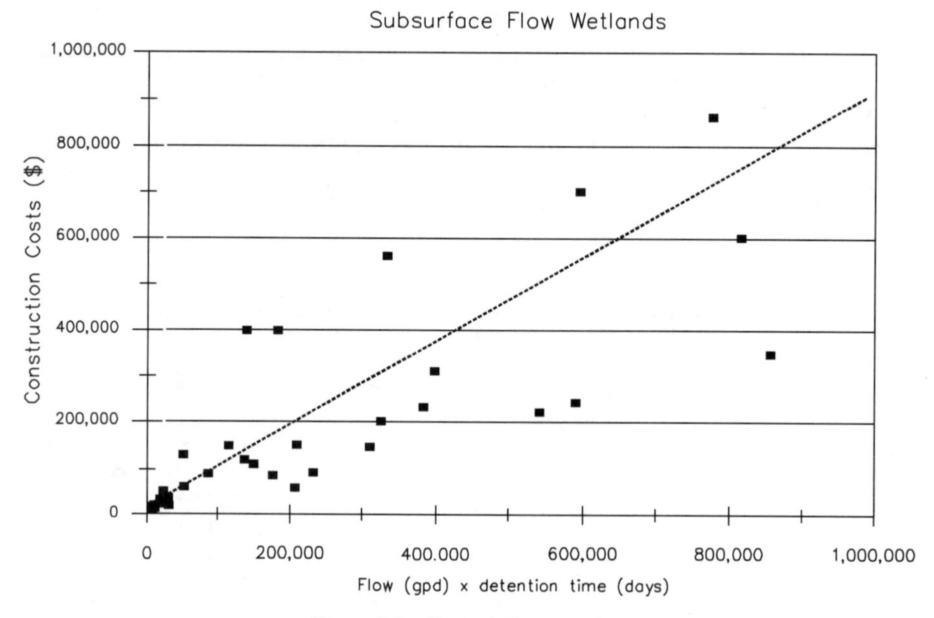

Figure 3.9 *Cost, dollars per day.*

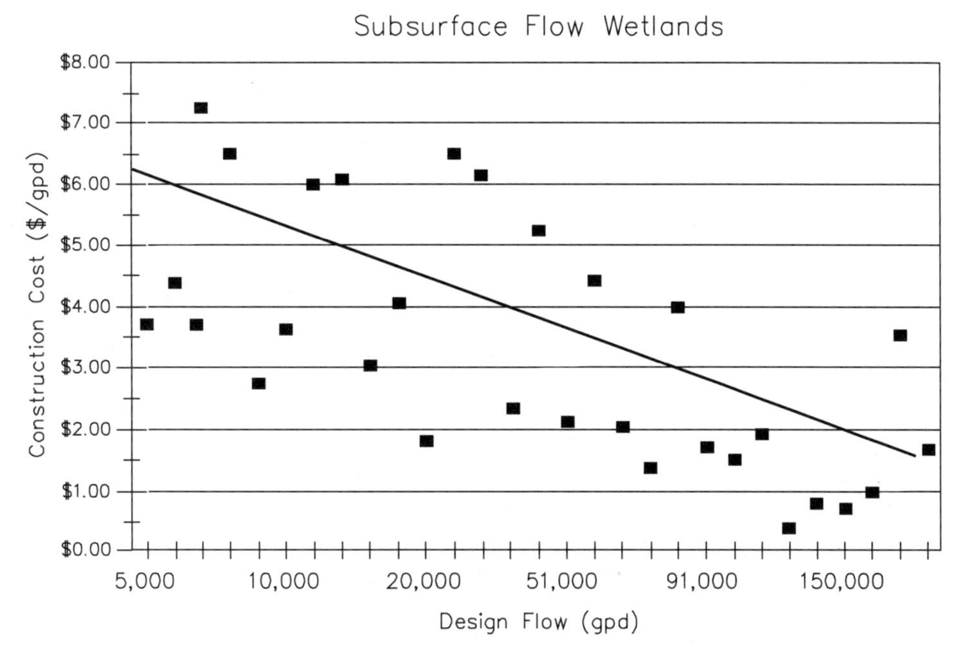

Figure 3.10 *Cost, gallons per day.*

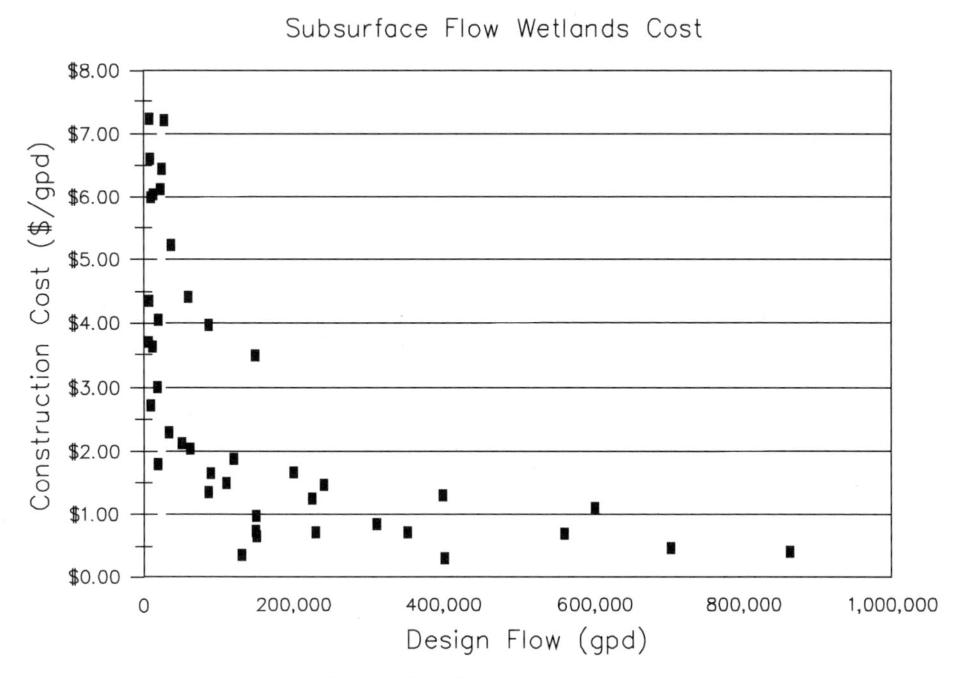

Figure 3.11 *Cost, gallons per day.*

wetlands labor costs are often lumped into the total system cost. However, costs can be estimated by inspecting the design and by recognizing that, in many respects, wetlands are very similar to wastewater stabilization lagoons from a maintenance and operational perspective. There are few items in wetlands systems that require maintenance or energy.

Operational costs can be divided into the following general categories:

Operation: testing, water level adjustment

Maintenance: weed control, flow distribution, and level adjustment sumps

Testing influent and effluent is generally the single largest cost. This cost will depend upon the frequency of testing, the number of water quality parameters, and the number of samples. For a BOD, TSS, total Kjeldahl nitrogen (TKN), and nitrate and ammonia sample, the costs will be approximately $150 per sample.

The level adjustment function usually does not require any attention. Water levels should be checked periodically (monthly or weekly on small systems, and daily on systems >100,000 gallons per day) to ensure that surfacing has not occurred in the subsurface flow system and that there is some flow through the system. In surface flow systems, the level adjustment must be checked for flow and the level in the wetlands must be visually inspected using a fixed gauge.

For the sake of appearance, weeds should be controlled around the edges, and large weeds should be removed from the gravel bed in the early spring. Plant debris in surface flow wetlands can be ignored as long as it does not affect the flow; for example, plant debris after a severe storm may blow downstream and clog the collection piping or level adjusting structures. Regular inspection of the flow distribution and collection devices should be part of the operating requirements for the system. Flow splitters using weirs should be checked and cleaned periodically.

Some systems have incorporated annual harvesting of wetlands plants. In the fall, before the plants become senescent, they are mowed and the litter is removed to a composting operation. The rationale for this operation is that it removes the stored nitrogen that would otherwise be released during the following spring. Although there is only a limited amount of information regarding this type of operation, the cost does not justify the value of the harvested nitrogen or the minimal increase in nitrogen removal. One benefit that many operators have noticed is that annual harvesting improves growth of new shoots in the spring.

Actual reported costs for all operations support the notion that wetlands are very inexpensive systems. The annual operation and maintenance costs for Denham Springs, Louisiana (3 mgd) were $29,550, which included the costs of operating the aerated lagoon and chlorinator. Mesquite, Nevada (0.4 mgd), has an operating budget of $10,000. This provides a range of 2.6 to 6.80 cents per1000 gallons per day.

Cost Comparisons with Other Technologies

Subsurface flow constructed wetlands have been used primarily for small-community systems. As the database indicates, over half of the systems are less than 100,000 gallons per day. The reasons for use as small-community systems are primarily economic as the case illustrated in Figure 3.12 demonstrates.

The small development of Hanson's Lake, Nebraska, has 235 homes located around two small lakes adjacent to the Platte River. The nearest sewer interceptor for the city of Omaha is 1.5 miles away. Design flow is 75,000 gallons per day. The options considered during planning were the following:

Pump to the City of Omaha
Mechanical package treatment system
Constructed wetlands and sand filter
Wastewater stabilization lagoon and sand filter

The city of Omaha has a wastewater treatment charge of $1.46 per 1,000 gallons which is typical for municipal systems without advanced wastewater treatment (AWT) standards. In addition, Omaha requires a $300,000 capital charge. This capital charge of approximately $1,300 per home is now typical of the charges many municipalities are initiating as part of the sewer hookup fee. Other capital costs include the cost of a lift station and 1.5 miles of sewer line to the city interceptor.

The mechanical package treatment system proposed was a sequencing batch reactor (SBR). SBRs have an excellent operating history and produce an excellent quality effluent. However, the operating costs are significantly higher than those of constructed wetlands or of treatment at Omaha. This is typical of small systems.

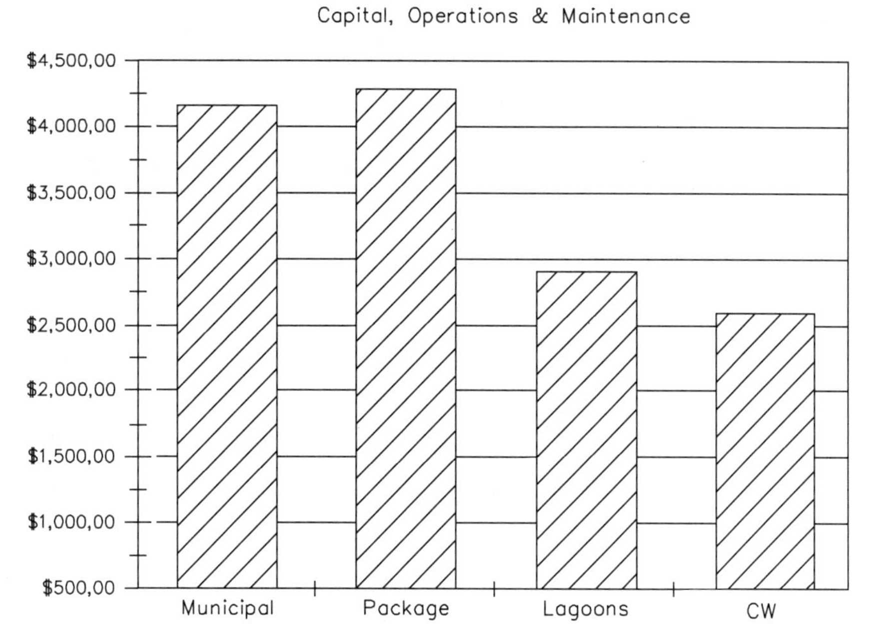

Figure 3.12 *Cost comparison.*

TABLE 3.6 Cost Comparison of Treatment Systems with a Twenty-Year Life Including Land, Legal and Engineering Costs

Option	Capital	O&M	Total
City of Omaha	$826,000	$1,420,253	$2,720,779
SBR	$596,700	$1,657,902	$2,683,889
Lagoon and sand filter	$742,600	$139,726	$1,416,984
Wetlands and sand filter	$365,300	$206,902	$835,012

Operation costs of small mechanical treatment facilities such as SBRs, oxidation ditches, and rotating biological contactors (RBCs) generally are higher than those of a large city wastewater facility simply because of the efficiencies of scale. It is not surprising that SBRs are expected to cost $2.50–$3.00 per 1,000 gallons.

Both wastewater stabilization lagoons and wetlands have nominal operating expenses. There is no equipment to maintain, no electrical costs, and both systems can treat a certain amount of sludge. Wastewater stabilization lagoons, although less expensive to build and operate, require 160 acres of land to provide a buffer for odors, and since land is not immediately adjacent, a small lift station would be required. The wetlands do not need the buffer; consequently, as Table 3.6 indicates, the wetlands is the least costly option. The twenty-year total is based on 3 percent inflation rate for operation and maintenance, and a 6 percent interest rate for the construction loan.

To further simplify the cost issues, consider the following example of a golf course subdivision:

120 homes, 60,000 gallons per day design flow with the following alternatives:

1. SBR, $250,000 construction cost, $2.50 per 1,000-gallon operating cost.
2. Constructed wetlands and sand filter, $265,000 construction cost, 10 cents per 1,000-gallon operating cost.

Although the constructed wetlands and sand filter cost slightly more, the significantly lower operating and maintenance costs make the constructed wetlands the most cost-effective choice. The difference in capital costs will be repaid in 104 days.

Summary of Construction and Operational Costs

Any energy-intensive wastewater treatment technology will always be much more expensive than constructed wetlands to operate. In general, subsurface wetlands will always be less expensive to build. The basic exchange is energy for land. As land for the treatment system increases via use of wetlands, the energy costs and capital costs decline. As land decreases, energy must be added to the wastewater treatment process to accomplish what natural processes accomplish without this assistance.

An obvious missing element in the cost structure is that the cost of land, except in the example of Hanson's Lakes. Since most wetlands have been built in rural areas where land costs are low, or on land that is not suitable for building, land costs generally have had little impact on the cost of constructed wetlands. However, if wetlands are considered for urban areas, then the cost/benefit analysis should include land costs and the benefits that accrue to the open space, habitat, and recreation.

Because of the simplicity of the technology, the low operating and construction costs, and the additional benefits that accrue (see the U.S. EPA publication *Constructed Wetlands for Wastewater Treatment and Wildlife Habitat: 17 Case Studies*), wetlands must be considered in the design of new wastewater treatment systems and the upgrade of existing systems.

Surface Flow Wetlands

The costs of surface flow wetlands are essentially the same as those of subsurface flow wetlands, except for the cost of the gravel. And as Table 3.7 shows, that can be a very significant factor.

Clearly, there is a cost savings for the use of surface flow constructed wetlands, and in general, the cost of a 2-acre system will approach $50,000 per acre; per acre costs will decline as the system becomes larger.

Even though subsurface flow constructed wetlands are more expensive as part of an overall design strategy, it is usually more cost effective to use subsurface flow gravel beds with anaerobic pretreatment for systems under 75,000 to 80,000 gpd and to change to aerobic pretreatment and surface flow wetlands. If your design falls in this range, both options should be examined.

THE REGULATORY PROCESS

If wetlands, as one part of a multipart treatment process, are selected, the designer must be prepared for another major hurdle: the state and federal regulatory processes. Although wetlands may be based on sound scientific principles, and the design principles described in standard engineering textbooks and EPA manuals, the likelihood that the regulatory agency to which you are submitting a permit will be receptive is very small. Most regulators are concerned only with the requirements of the regulations that they are charged with enforcing, not with the cost to the community or the possibility that there may be new and better methods of

TABLE 3.7 Cost of a Typical 50,000 Square-Foot Surface Flow Constructed Wetland

Description	Unit	Unit Price	Total $	Percent
Excavation/compaction	cubic yard	$1.75	13,000	19.4
Soil—18 inches	cubic yard	$1.00	2,800	4.2
Liner—30 mil PC	square foot	$0.35	19,250	28.7
Plants—24 inch o.c.	each	$0.60	7,500	11.2
Plumbing	lump sum		7,500	11.2
Control structures	lump sum		7,000	10.4
Other	lump sum		10,000	14.9
	Total		$67,050	

treatment. Regulators are usually gatekeepers, not problem solvers. They have been hired to let pass wastewater treatment technologies previously approved by their superiors. Most regulatory agencies work off a set of "cookbooks" or design-based standards describing acceptable treatment elements. This type of system exists because most state regulatory agencies cannot afford to hire trained engineers to review designs and approve new technology. If your design is not in the cookbook, it will be extremely difficult to gain approval. This process is similar to saying that all cars in the state will have four doors, V-8 engines, and 15-inch diameter wheels.

Possibly the worst example of this approach is the Indian Health Service, which is responsible for the design, approval, and funding of wastewater treatment facilities on many Indian reservations. Engineers of the Indian Health Service have determined that the only suitable technology for use on the reservation is wastewater stabilization lagoons, a technology whose formal development dates from the 1920s. Scarce water resources are retained in these lagoons and are allowed to evaporate rather than be reused. Reuse would require a different technology.

Many regulatory agencies rely on regulations based on 15 year old technology. Since regulators must have at their disposal many different technologies, regulatory guidelines are hundreds of pages long. The review process is unimaginably complex to most people as a consequence, and also very time-consuming. Imagine what would happen to our society if every automobile—and I mean *every* automobile, not just the model—made in Detroit were subject to a design review in which a nonautomotive engineer checked to see if the parts used were acceptable and gave you approval to drive the car only after such a review? Can you imagine what would have happened to the computer industry if it had required product approval by a regulatory agency using technology guidelines that were always at least fifteen years out of date?

Fortunately, there is another alternative that offers a better method of dealing with technological innovation and regulatory review. This is the discharge permit, which simply says that the water leaving the wastewater treatment facility shall meet certain standards. It doesn't specify how treatment should be done. As a consequence, it is possible to write regulations that are three pages long. The community representative requesting a discharge permit simply appears before the regulatory agency and is told what the standards shall be for the right to discharge treated effluent. The agency then simply becomes a policeman and issues tickets for "speeding," that is, exceeding the limits imposed by the permit.

It is important to understand that the regulatory constraints have a significant economic impact, whether they are technology based or performance based. And when the community attempts to use a new approach, the regulatory hurdles may appear daunting or, worse, insurmountable. The fault may lie in jurisdictional issues or in the inability of the regulatory agency to make sense of conflicting regulations. The following two examples are not uncommon experiences.

LaGrange County had a particularly frustrating permit experience, and not because of the intentions of the regulators. Remember that part of the county consists of a lake district where there are several thousand homes on septic tanks and leach fields built around many small lakes. When the residents of this district recognized that they were polluting the lakes and endangering their health, they decided to install sewers and build a wastewater treatment system. Note that there was no requirement to do so. When the Indiana Department of Environmental Management (IDEM) was presented with the initial proposal to design a system using wetlands and infiltration basins (which treat wastewater by filtration and denitrification), they responded that they did not have the administrative authority to permit discharges to the ground. Therefore, the county would have to obtain discharge permits for discharge to the streams that feed the lakes.

This is a classic case in which the regulating agency is stuck with regulations that place the community in a Catch-22 situation. Clearly, the best solution is to apply the treated effluent with the remaining nutrients of nitrogen and phosphorus to the land where it can be utilized by crops, prairies, or forests, instead of letting it run into the lakes and streams, where the nutrients result in algal blooms and repeat the existing problem.

In New Mexico, a developer had sixty lots that were big enough to allow construction of individual on-site septic tanks and leach fields. These require no monitoring or testing and are notorious for discharging high levels of nitrogen into the groundwater. The developer instead opted to upgrade the treatment by first providing sewers to each lot and then adding an additional treatment step using constructed wetlands. Because the consolidation placed him under a different set of regulations, the homeowners are now under a more onerous set of regulations that require frequent testing, monitoring and reporting. Although the costs are not significant, amounting to about $3 per month per household, they represent a regulatory process that penalizes the developer rather than rewards him for his efforts in improving the water quality.

Arizona offers another daunting experience for innovators. In a meeting with regulators, the engineer and developer described a design approach that would have made a significant improvement in conventional approaches by using a low-energy system that included landscape and habitat development. As the meeting was ending, the engineer asked the head of the technical review team how long the permit would take. The reviewer responded that the proposed technology (constructed wetlands) couldn't be permitted because there were no design guidelines. When asked how long he thought it would take to develop those guidelines, the regulator responded that they had been working on swimming pool guidelines for six or seven years and didn't have any yet.

Often several permits and jurisdictions are involved. An extreme example, presented in Figure 3.13, shows the permit procedure for a project in Maricopa County, Arizona. In addition to multiple levels of review within the regulatory agency, there are several different permits. For example, the Arizona Department of Environmental Quality requires an Aquifer Protection Permit, a Groundwater Recharge Permit, and a Reuse Permit. The state agency states that the permit process takes 10–12 months. It is not uncommon for wastewater discharge permits to take longer than planning and zoning permits. The client should be informed of the permitting process at the beginning of the project. The designer should meet with the regulatory agency(ies) and the owner at a preliminary meeting to discuss permitting issues.

This section has tried to demonstrate that the development and implementation of a community wastewater treatment facility requires a multidisciplinary approach involving many highly specialized professionals. Ultimately, however, it is the re-

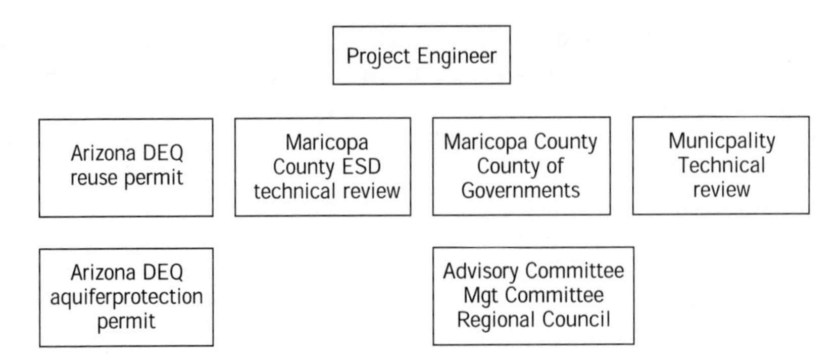

Figure 3.13 *The permit process.*

sponsibility of the community to educate itself prior to seeking a final solution and direct the professionals accordingly. The community must also recognize that the regulatory process can limit their options if allowed to do so.

The regulatory process is not required to be rational or evolutionary. In general, experiences of other states are not considered relevant and are often looked upon as mistaken or ill considered unless they support current practices. U.S. EPA publications and recommendations are sometimes pointedly ignored. As an example, the State of Indiana was encouraged by the EPA to develop design guidelines for the subsurface disposal of groundwater. Encouragement was provided by grants given to two small communities to use disposal fields based on design guidelines published in the U.S. EPA *On-site Wastewater Treatment and Disposal Systems* design manual. These systems were designed, permitted, and built in 1988 against the express wishes of the regulator in charge of reviewing facility plans. The technology used was neither new nor experimental and is routinely used by adjacent states. As of 1998, the State of Indiana does not have any regulations regarding the subsurface disposal of treated wastewater even though 52 percent of the population currently uses septic tanks and leach fields.

Regulators generally do not have any economic restraints placed on their interpretation of regulations. Often they interpret their role as defining a standard to be met by the permit applicant using technology that is acceptable to the regulator. In effect, the discharge standards are a hurdle set by the regulatory agency that must be cleared, independent of cost, community size, or potential risk to health. As an example of a health risk, nitrate concentrations in groundwater were set at 10 mg/L by the U.S. EPA and subsequently adopted by most states, although the original scientific recommendation was 45 mg/L. Some states, such as New Mexico, and Arizona, then regulated the total nitrogen concentration and have set the standard as 10 mg/L, even though there is no evidence that all the nitrogen is converted to nitrates.

It is essential that whenever a permit is submitted for constructed wetlands, or for any new technology, the designer must have a comprehensive package including the following:

Permit application
Plans and specifications
Flow schematic drawing

Process flow description and design calculations

All appropriate reference materials demonstrating the efficacy of the proposed technology.

In addition, the designer should be prepared for many hours of educational sessions. When it comes to new and more cost-effective technology, most regulatory agencies have no incentive to permit its use. Although many regulators recognize the need for improvements, the rules for many states are the prescriptive standards described previously. "If it isn't in our book it can't be done" is unfortunately all too common.

Hopefully, you will not be the first engineer in your state to handle a wetlands project, and you will be meeting with an informed and knowledgeable regulatory agency that is familiar with the textbooks and design manuals that are the basis of your project. In any event, be prepared.

4

Design, Operation and Maintenance of Constructed Wetlands

Nature to be commanded, must be obeyed.

—Francis Bacon (*Novum Organum,* 1620)

Knowledge is power.

—Francis Bacon (*Meditationes Sacrae,* 1597)

This chapter is intended to provide the necessary practical and theoretical information so that the engineer and designer will be able to understand fully the problems associated with designing constructed wetlands. The importance of shape and hydraulic conductivity is not well understood by many designers. Gravel size, distribution, and sieve analysis, as well as the importance of the distribution of wastewater, are often overlooked in more technical treatises on constructed wetlands (Kadlec and Knight, 1996; Reed, 1993). These practical aspects of design have often been misunderstood, and many examples of wetlands are not performing well because the designer has not paid attention to gravel specifications and hydraulic conductivity.

DESIGN PRINCIPLES

Constructed wetlands can be designed to remove both soluble and solid organic compounds (BOD), TSS, nitrogen, phosphorus, metals, hydrocarbons, and organic priority pollutants, as well as pathogenic bacteria and viruses. The design principles are based on hydraulic capacity, residence time, areal loading rates, water temperature, and plant density. These concepts can be reduced to the commonsense idea that a wetlands, depending upon its size, can process so many gallons of water per day, and so many pounds of fertilizer and organic material. The latter quantities must vary seasonally with more fertilizer (nitrogen and phosphorus) being processed during the peak growing season.

In fact, the processes by which the various elements that make up the pollutants of concern are very complex. Figures 3.1 and 3.2 show the carbon and nitrogen pathways respectively, in a wetlands environment. Hundreds of species of bacteria and other microorganisms are involved, as well as plants and higher animals (Mitsch & Gosselink, 1992). The microbial interactions are extremely complex and very difficult to model with any consistency (Stolp, 1988). Rather than attempting to model these complex processes, engineers have resorted to treating the wetlands as a black box. By making large number of measurements of influent and effluent water quality, they have been able to arrive at simple formulas that describe these processes with sufficient accuracy. Consider Figure 4.1, a black box for carbon removal (BOD).

As the figure indicates, there are only three pathways for carbon: soluble and solid compounds in the effluent, gaseous compounds (e.g., CO_2 and CH_4) returned to the atmosphere, or solids retained in the wetlands. In fact, there are no measurements for the amount of carbon returned to the atmosphere, but the measurements of carbon in the influent and effluent are really all that we need to know to predict performance.

Accumulation of solids has different effects in the two types of wetlands. Some carbon compounds are converted to plant materials, bacteria, and other microorganisms and form the basis of the food chain. For example, cattails are eaten by grasshoppers, which in turn become food for sparrow hawks. Plants and bacteria grow and die, accumulating in the bottom of surface wetlands as reduced organic compounds, eventually becoming peat and then coal. In subsurface flow wetlands, plant litter accumulates on the surface of the gravel bed, developing into a compost material similar to leaf litter. Organic solids also accumulate in the interstices of the gravel as refractory organic compounds (long chain organic molecules with few oxygen molecules). Various estimates have been made regarding the time it takes for the solids to fill the voids. Only two wetlands have been tested in the TVA region (Hines, 1996) and the results show 33 to 150 years before the gravel would have to be replaced. Since the solids accumulate more readily in the front of the gravel bed, only the front third of gravel will need to be replaced.

The nitrogen pathways are perhaps even more complex, and seasonal effects are more pronounced. Ammonia removal is not just a temperature-dependent phenomena; it is also seasonal. Vegetation that died off in the fall and accumulated over the winter releases ammonia in the spring as water temperature and bacterial activity increase. Instead of metabolizing this ammonia, the wetlands release it in the spring. Graph 4.1 shows both the spring spikes and the temperature effects.

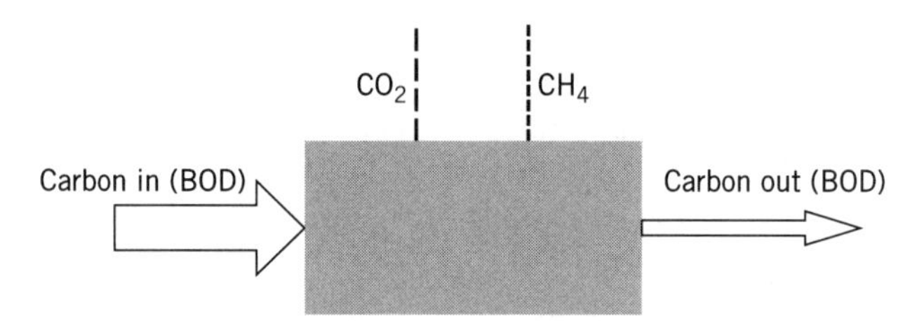

Figure 4.1 *Black box.*

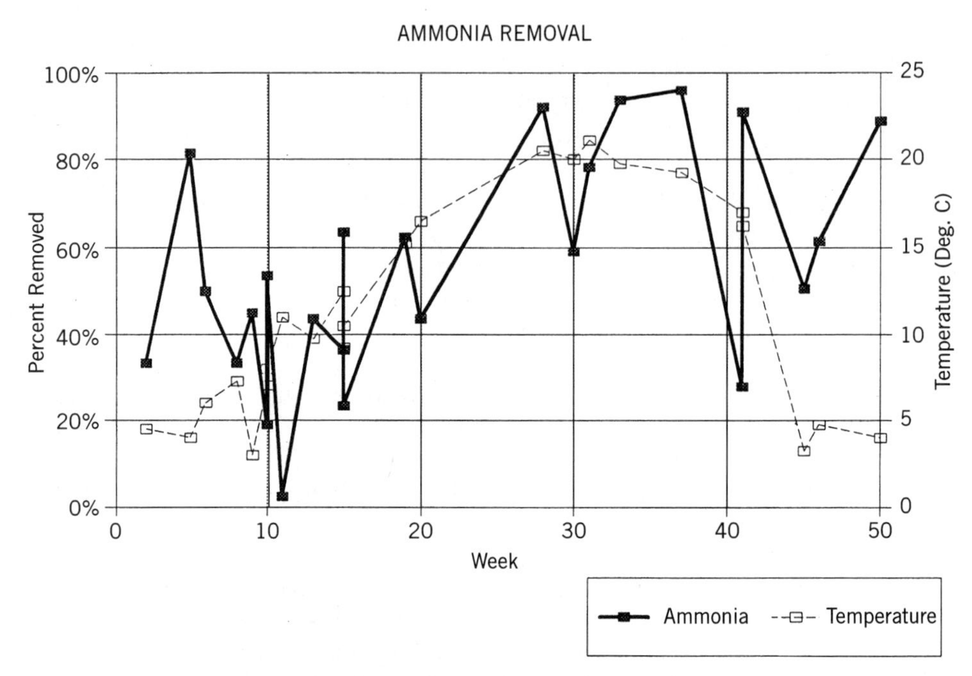

Graph 4.1 *Seasonal ammonia removal.*

Hydraulics

The hydraulic capacity of a wetland can be defined as the ability of the wetland to process a given volume of water in a given time. Thus, a constructed wetland can be said to be capable of treating, for example, 10,000 gallons per day with a hydraulic residence time (HRT) of 3.5 days. The HRT time is the expected average time in which a molecule of water will flow from one end of the wetland to the other. In natural systems this will vary greatly from season to season and from year to year because of fluctuations in rainfall in the drainage basin. In natural wetlands, the area occupied and the depth will fluctuate with the season. This area fluctuation is acceptable when designing for the treatment of storm runoff, but when treating wastewater, the area of the treatment facility and the depth of the water are usually controlled with hydraulic structures that ensure that wastewater flows are uniformly distributed over the entire constructed wetland. Thus the HRT can be calculated with some accuracy. This calculation is important in defining expected treatment results.

The problem of predicting the time it takes for water to flow through a dense stand of cattails and bulrushes in a surface flow constructed wetland or through a subsurface flow constructed wetland (a gravel bed planted with reeds and cattails) can be dealt with by using two different methods. Before considering these techniques, a brief discussion of the importance of geometry in designing constructed wetlands is necessary.

Early designs modeled natural water courses that tend to be long and narrow. When treating wastewater, this presented some problems. Organic solids settled out in the front of surface flow wetlands in such concentrations as to cause odors. In the case of subsurface flow wetlands, the front end of the gravel bed became clogged with solids, and the wastewater began to flow on the surface. This usually affected residence times and therefore treatment. The reasons for these problems can be illustrated with the following two figures.

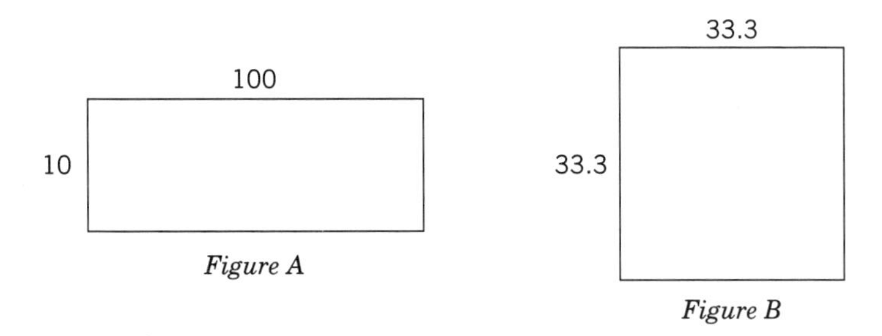

Figure A

Figure B

Figure A measuring 10 feet by 100 feet and Figure B measuring 33.3 feet by 33.3 feet, both have areas of 1000 square feet. If both figures have a depth of 2 feet, then their total volumes equal 2,000 cubic feet (cf). If 200 cubic feet per day (approximately 1,500 gallons) is introduced into each wetlands then in 10 days the wetlands will fill up and begin overflowing. As water is continuously introduced at the rate of 200 cubic feet per day on day eleven, 200 cubic feet will leave each wetland. Both systems have the same (HRT) and hydraulic loading rate (HLR). HRT is based on the total volume of the wetland and HLR is based on the area. If the depth is decreased, then the areas will increase and the HLR will decline.

$$\text{HRT} = \text{volume/flow} = 2000 \text{ cf}/200 \text{ cfd} = 10 \text{ days}$$

$$\text{HLR} = \text{flow/area} = 200 \text{ cfd}/1{,}000 \text{ sf} = .2 \text{ cf/sf/day} = .2 \text{ ft/day}$$

However, note that in Figure A, the 200 cubic feet must travel 100 feet in 10 days, or 10 feet per day while in Figure B the 200 cubic feet travels only 33.3 feet in 10 days, or 3.3 feet per day. Thus, in Figure B, the velocity of flow is one-third as great. This is extremely important in the removal of solids from wastewater. The slower the velocity of the water with suspended solids, the more likely that the solids will settle out. Thus, Figure B is three times more efficient in settling solids.

Perhaps more important is the vertical cross-sectional area (Avs) at right angles to the direction of flow of each figure. For each figure

$$\text{Avs} = \text{depth} * \text{width};$$

$$\text{Figure A Avs} = 2 * 10 = 20 \text{ square feet.}$$

$$\text{Figure B Avs} = 2 * 33.3 = 66.6 \text{ square feet.}$$

Figure B's Avs is 3.33 times larger and therefore three times more capable of handling solids. In our example, with each figure receiving 200 cf per day, each figure has the following vertical HLR:

Figure A Vertical HLR = 200 cf/20 sf = 10 cf/sf/day

Figure B Vertical HLR = 200 cf/66.6 sf = 3.33 cf/sf/day

Figure B is receiving only one-third the solids per square foot of vertical cross section. Thus, it is not hard to understand that Figure A is more likely to experience clogging of the gravel bed in a subsurface flow constructed wetlands, or to accumulate solids faster than can be digested aerobically in a surface flow wetlands.

From the previous discussion, it is clear that short, widely constructed wetlands will perform better than long, skinny ones. But how much shorter? And how does the designer avoid preferential flow paths that are to be expected in short, wide constructed wetlands. Consider Figure A again, and this time rotate it 90 degrees so that the flow of water now travels only 10 feet.

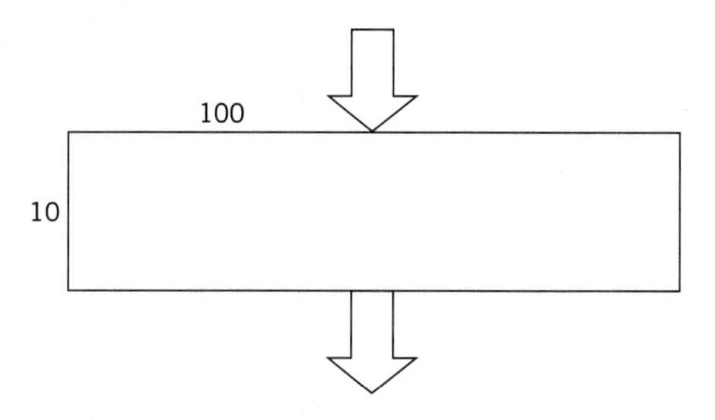

The problem for the designer is to ensure that the flow is evenly distributed. This can be accomplished by installing a 100-foot-long weir along the influent end, dividing the wetlands into 10 cells, and collecting the effluent in another overflow weir.

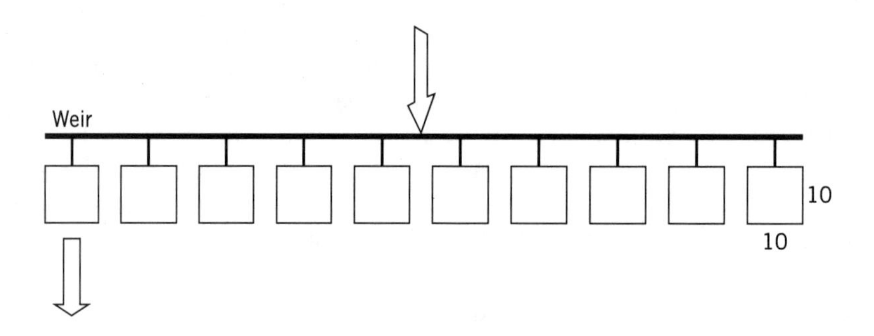

This figure can now be divided into ten separate cells, disconnected and placed in ten different locations in the landscape. Each cell can be constructed as needed. This particular approach is very useful when designing for a new subdivision or for fast-growing communities. Cells are built as needed.

Next, let us complicate the idealized situation described above by adding plants and gravel. In the case of a surface flow wetland, the actual volume of water will be somewhat less than the 2,000 cubic feet because of the mature mass of the plant stalks and leaves. The actual percentage of water displaced by the plants is estimated to be approximately 25 percent. Thus, in our example, the actual volume of water and HRT are:

$$\text{Volume} = 2000 * (1 - .25) = 1{,}500 \text{ cubic feet}$$

$$\text{HRT} = 1{,}500/200 = 7.5 \text{ days}$$

The mature mass of the plants will therefore make a significant difference in the HRT, and should be accounted for when designing surface flow wetlands. Notice how the velocity changes as well.

With gravel, the problem is complicated by the size of the gravel and the distribution of the different gravel sizes. Generally, as the gravel size gets larger, the void size increases but the void ratio (i.e., total volume of voids/total volume of rock and voids) decreases. Larger rock has smaller total voids but greater hydraulic conductivity. For example, sand has a greater total void volume than pea gravel but lower hydraulic conductivity. Figure 4.2 shows this relationship for a particular gravel pit.

Equally important is the distribution. Gravel with a particular distribution, say 100 percent passing a 1-inch sieve and 100 percent retained by a 0.5 inch sieve

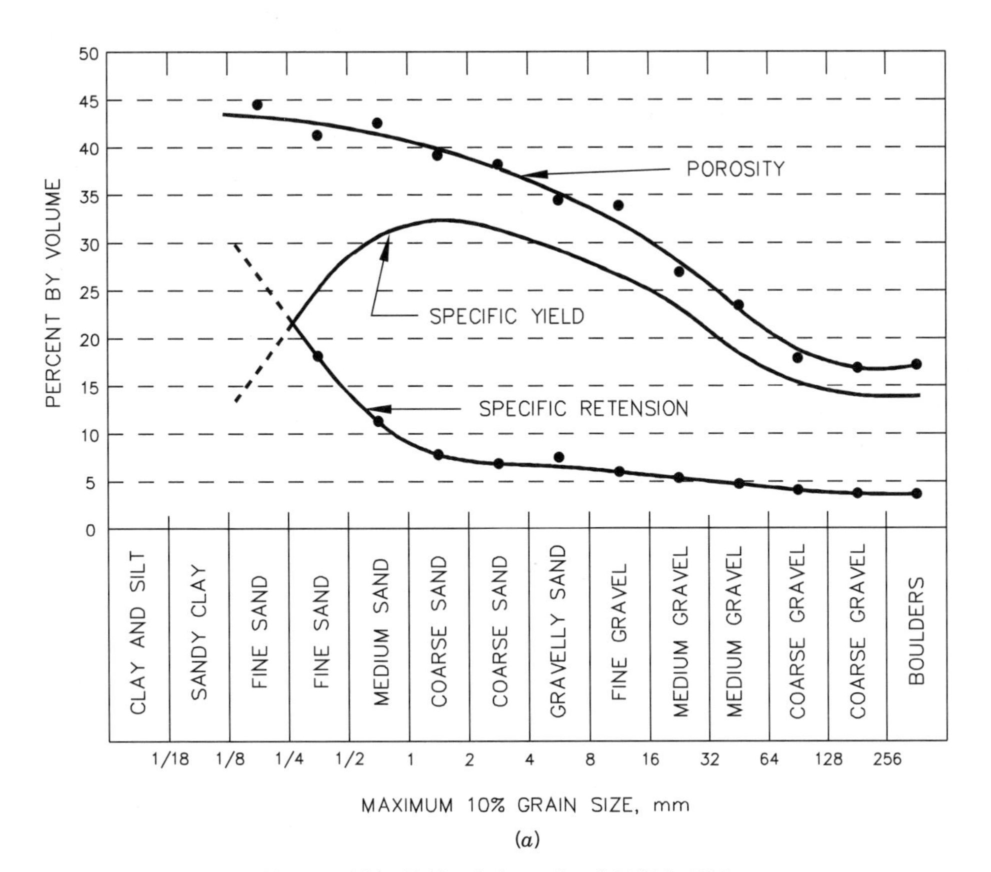

(*a*)

Figure 4.2(a) *Void ratio/porosity, yield (U.S. EPA).*

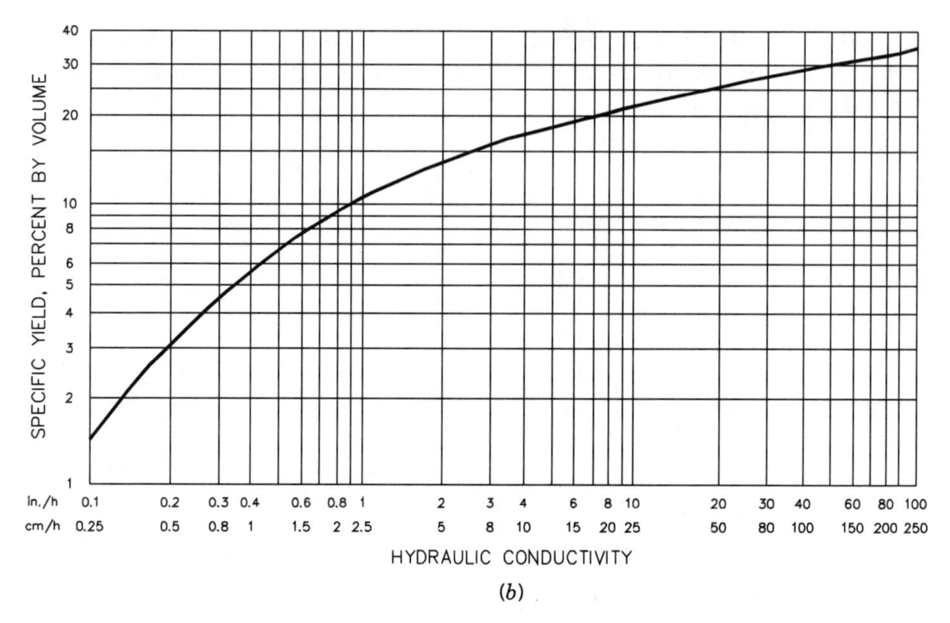

Figure 4.2(b) *Yield versus conductivity (U.S. EPA).*

will have a very different void ratio and hydraulic conductivity than gravel with 100 percent distribution passing the 1-inch sieve. This latter gravel will have more fines which fill the voids.

The problems associated with the uniform distribution of water are part of the engineering design process. Solutions are as old as the art of irrigation in the Fertile Crescent of the Middle East. Gravity flow solutions are preferable to solutions using pumps. Note that in the preceding discussions, all of the examples are rectangles and squares. Generally, these shapes are not useful to the landscape designer, nor are they found in natural wetlands. More natural shapes can be incorporated into the design by placing rectangles and squares within irregular shapes. Figure 4.3 illustrates how this might be accomplished.

The sections on materials and the construction process allow for the designer to understand how the more natural shapes can be incorporated into the design without inordinate expense.

Temperature

Except for the settling of solids, all wastewater treatment processes are temperature-dependent biological processes. Ever since the Dutch biologist Arrhenius described bacterial growth processes in 1742 as temperature-dependent processes, wastewater treatment processes have used this basic observation to describe the treatment process. In wastewater treatment processes, and especially in wetlands, the problem is to determine the water temperature. A simple description of a heat model for wetlands follows:

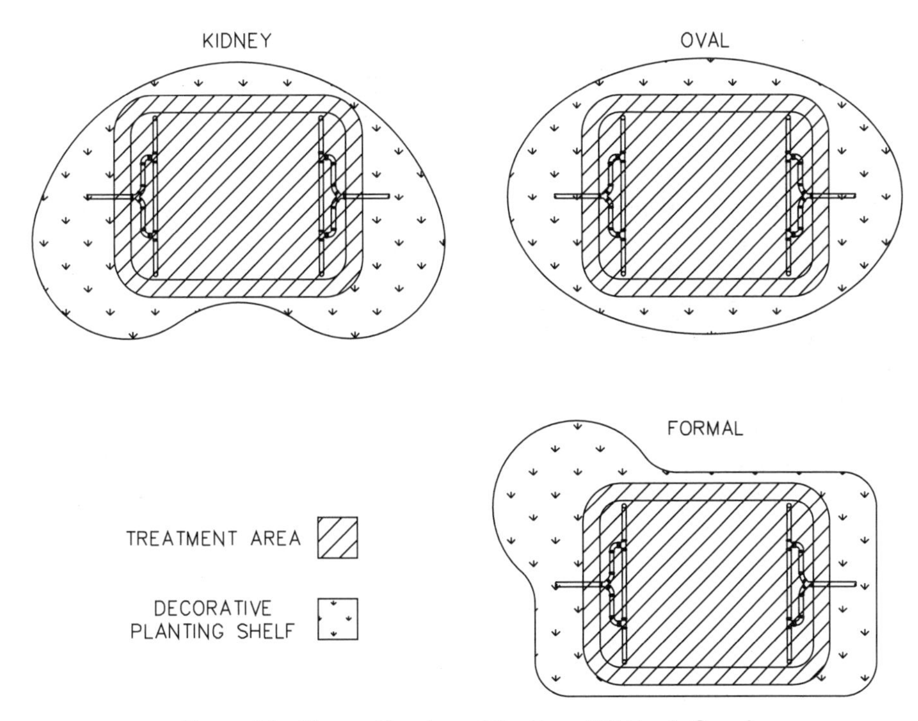

Figure 4.3 *Shapes (Courtesy of Southwest Wetlands Group).*

$$\text{delta heat} = \text{heat(in)} \pm \text{heat (air)} \pm \text{heat(ground)} - \text{heat (out)}$$

where

delta Heat = the change (in BTUs or calories)
heat(in) = heat content of the influent water
heat(air) = heat lost or gained to the air
heat(ground) = heat lost or gained from the ground under the wetlands
heat(out) = heat content of the effluent water.

The problem is complicated in determining heat transfer coefficients for losses to the air and ground. Layers of dead vegetation, snow cover, and ice obviously affect heat transfer to the atmosphere, so various coefficients have been proposed by Sherwood C. Reed in his excellent engineering text *Natural Systems for Waste Management and Treatment,* A simplified version for heat loss calculations can be accomplished using the following equation:

$$\text{heat loss/gain to air} = \text{surface area} * \text{Uair} * (\text{Twater} - \text{Tair}) * \text{HRT}$$

where surface area is the surface area of the wetlands and

U air = 0.052 BTU/ft^2 − °F − hr
Twater = temperature of water entering the wetlands (°F)
Tair = air temperature (°F)
HRT = hydraulic residence time (hours)

Notice that the area of the wetlands is required to perform this calculation, and since we do not know the area, some assumptions have to be made about water temperatures and then checked after checking the area. Calculation of area and temperatures will be an iterative process.

Geometry also plays a role in heat loss. Consider the design examples used in the hydraulic examples above. The perimeter of a wetland with a width of 10 feet and a length of 100 feet is 220 feet. A wetlands 33.3 feet square has a perimeter of 133.2 feet. Like the envelope of a house, the larger perimeter will lose more heat. Although this is an intuitive concept, current design models for heat losses and gains do not take into account these edge effects. Clearly, they are important.

Design Goals

Now that we have figured how to get the water where we want it, we now need to establish our wastewater treatment goals. The main constituents of wastewater that the designer will be concerned with removing or reducing are:

Biological oxygen demand (BOD)—a measure of the organic compounds in wastewater that require oxidation in order to become stable
Suspended solids (SS)—the organic and inorganic particles in wastewater
Nitrogen—a pollutant if discharged into rivers; a resource if used to irrigate crops or the landscape or to create a habitat

In these definitions, it is clear that in some circumstances the pollutants are not necessarily pollutants. A basic principle of design therefore is to consider the options for disposing of or reusing the treated effluent. The options are:

Discharge to rivers, streams, lakes, ponds, or the ocean
Discharge to the ground
Land application—irrigation, habitat creation
Evaporation

Each of these options will have different standards for BOD, SS, and nitrogen. The higher the percentage of pollutants that are removed, the greater the cost. In general, land application and evaporation have the easiest permit requirements. In the case of evaporation, which is often used in the arid West, there are no standards. In the case of land application, the pollutants, especially nitrogen, become resources. As mentioned in Chapter 3, wetlands have advantages and limitations. How, then, can we design our wastewater treatment facility using wetlands as a treatment element?

SUBSURFACE FLOW CONSTRUCTED WETLANDS

To understand the design formulas, think of a wetland as a big field that has certain nutritional requirements. Plants need carbon, nitrogen, phosphorus, sulfur, calcium, potassium, sodium, and a host of micronutrients, all of which are found in wastewater. So, the questions that the engineer must answer are: How much of these elements does this wetland require for its growth? What happens to these elements during different times of the year? How much is discharged downstream or to the atmosphere or stored in the decayed plant mass? All of these questions have been answered by observing constructed wetlands over many years and measuring the amount of material that entered the wetlands, and the amount that left.

An obvious problem that must be dealt with is the differing rates of plant growth. Clearly, there is no plant growth in the winter in most parts of the United States, and microbial activity must also be at a minimum. So, there is obvious seasonal activity that makes it desirable to apply wastewater during the growing season and store it during the winter. However, often this is not practical. More fundamentally, we know that all wastewater treatment processes are primarily microbial, and this activity is a temperature-dependent process. Hence we can write our design equation as follows:

$$\text{As} = \frac{Q * (\ell n\ Co - \ell n\ Ce)}{K_t * d * n}$$

where

Q = flow, in m^3/day
Co = influent BOD (mg/L)
Ce = effluent BOD (mg/L)
K_t = temperature-dependent rate constant
d = depth of gravel bed
n = porosity of gravel

To adjust K_t for subsurface flow wetlands, we need to make the following calculation:

$$K_t = 1.104 * (1.06)^{(T-20)}$$

where T is the temperature of the water in °C.

Example: Determine the area of a constructed wetland for a community of 100 homes, each with a septic tank. Assume that we will collect all of the wastewater, using the existing septic tanks as our pretreatment tanks.

Step 1. Calculate the flow. The U.S. Census tells us that the average number of people per dwelling is 3.2 and the U.S. EPA tells us that the average per capita flow is 50 gallons per day.

$$100 \text{ homes} * 3.2 \text{ people} * 50 \text{ gpdpc} = 16{,}000 \text{ gpd} = 60.5 \text{ m}^3/\text{day}$$

Step 2. Assume a water temperature. Septic tank effluent averages 16°C, but it will cool off by the time it gets to the treatment site. Assume that the water temperature is the same as the ground temperature, i.e., 10°C.

Step 3. Calculate the rate constant.

$$K_t = 1.104 * 1.06^{(10-20)} = 0.62/\text{day}$$

Step 4. Calculate the area: Assume that the influent BOD is 140 mg/L and that we would like the effluent BOD to be 10 mg/L. Assume a depth of 2 feet (0.6 m) and a porosity of 0.4.

$$\text{As} = \frac{60.5 * \{\ln(140) - \ln(10)\}}{0.62 * 0.6 * 0.4} = \frac{160}{0.145} = 1{,}103 \text{ m}^2$$

After the area is calculated, the temperature should be checked to ensure that the assumption regarding the temperature was reasonably accurate.

Another approach to the problem, which is much easier, and provides a good preliminary estimate, is to assume that the wetland can remove so many kilograms (pounds) of BOD per year in a temperate climate. A good estimate is 2.5 kg/m²/year (0.5 lb/ft²/year).

Example: Using information from above example,

Step 1. Calculate BOD removed per year.

$$\{(140 - 10) * 60.5/1000\} * 365 = 2{,}871 \text{ kg/yr}$$

Step 2. Calculate the area.

$$\text{As} = 2{,}871/2.5 = 1{,}148 \text{ m}^2$$

This is a good method for preliminary estimates and can easily be adapted to warmer climates. If the water temperature is 15°C, then the value becomes 3.3 kg/m²/year. For 20°C, the value becomes 4.4 kg/m²/year.

TSS is determined differently because the process for removing solids essentially involves filtration and retention time. TSS removal is affected by velocity (refer to discussion on geometry above). The equation for solids removal is:

$$\text{TSS}_{\text{eff}} = \text{TSS}_{\text{inf}} * (0.1058 + 0.0011 * \text{HLR})$$

where

HLR = hydraulic loading rate (cm/day)
TSS_{eff} = effluent TSS in mg/L

TSS_{inf} = influent TSS in mg/L

The HLR is calculated as follows using the above example.

$$HLR = Q/As = 60.5\ m^3/1{,}103m^2 = 0.055\ m/day = 5.5\ cm/day$$

If the influent TSS is 100 mg/L, then

$$TSS_{eff.} = 100 * (.1058 + .0011 * 5.5) = 11\ mg/L$$

Nitrogen removal is also a temperature-dependent process and is highly sensitive to cold temperatures. Once wintertime water temperatures drop below 5°C, nitrogen removal is problematic. The form of the nitrogen is also important. It is much easier for wetlands to remove nitrates than ammonia; hence, if nitrogen removal is a goal, then the treatment process should provide for nitrification, with subsequent discharge into a wetland for denitrification. The following formula calculates nitrogen removal:

$$\ell n\ (TKN/NH_{4\ eff}) = K_t * HRT$$

or

$$HRT = \ell n(TKN/NH4)/K_t$$

where

TKN = influent Kjeldahl nitrogen in mg/L
$NH_{4\ eff}$ = ammonia concentration in the effluent in mg/L
$KNH = 0.01854 + 0.3922(rz)^{2.6077}$
rz = percent of bed depth occupied by roots (decimal 0–1)
$K_t = KNH * (1.048)^{(T-20)}$
HRT = hydraulic residence time in days

For our typical temperate winter climate and water temperatures of 5–10°C, the range of values for K_t is 0.2 to 0.25.

$$\ell n\ (NO_{3\ inf}/NO_{3\ eff}) = K_t * HRT$$

or

$$HRT = [\ell n(NO_{3\ inf}/NO_{3\ eff})]/K_t$$

where

$K_t = 1.15^{(T-20)}$
$NO_{3\ inf}$ = influent nitrate in mg/L
$NO_{3\ eff}$ = effluent nitrate in mg/L

For our typical winter climate and water temperatures of 5–10°C, the range of values for K_t is 0.12 to 0.25. Denitrification is much more sensitive to cold than is nitrification; however, above 10°C, denitrification proceeds much faster than ammonia removal. At 15°C, the deintrification rate constant is 0.49 versus 0.31 for ammonia.

Example: Calculate HRT for ammonia removal wetlands, assuming that influent TKN is 45 mg/L and the permit calls for 4 mg/L in the discharge. If the water temperature is 5°C, then:

$$\mathrm{HRT} = \ln\,[(45/4)]/K_t$$

where

K_t = 0.2 (see above)
HRT = 12 days = volume/Q = (As * d * n)/Q

where

Q = flow, in cubic meters per day
d = depth in meters
n = porosity of gravel expressed as decimal fraction

The surface area, As = Q * HRT/(d * n) in our example is

$$\mathrm{As} = 60.5 * 12/(0.6 * 0.4) = 3{,}025\ \mathrm{m}^2$$

or almost three times the area required for BOD removal. The temperature assumptions will have to be checked against this calculation, especially with such a long HRT. With such a large area required for nitrogen removal, heat losses will be much greater than for BOD removal. Heat loss calculations should be redone based on this larger area and the iteration repeated until the temperature difference is calculated to within 0.5°C.

SURFACE FLOW WETLANDS

The design problem is very similar to that of subsurface flow wetlands, except that the reaction rate constants are different, and the designer should make provisions for ice cover during the winter. The same formula used to calculate the area for BOD removal can be used for surface flow wetlands, except that the reaction rate constant $K_t = .278 * (1.06)^{(T-20)}$. The porosity n is the density of the plant stems which typically in a mature wetlands are in the range of .65 to .75. Water depth should not exceed 0.45 m (18 inches). Calculations can proceed as in the previous example for subsurface flow wetlands.

The formula for TSS removal is:

$$\mathrm{TSS}_{\mathrm{eff}} = \mathrm{TSS}_{\mathrm{inf}} * [.1139 + .002\ (\mathrm{HLR})]$$

The nitrogen removal formulas are the same as for subsurface flow, except that the reaction rate constants are different. They are:

$$\text{Ammonia } K_t = .2187 * (1.048)^{(T-20)}$$

$$\text{Nitrate } K_t = 1.15^{(T-20)}$$

Note that the nitrate removal constants are the same for both surface and subsurface flow wetlands. See the spreadsheet in the Appendix for an example of the calculations for surface flow wetlands.

THE CONSTRUCTION PROCESS

To construct a wetlands, three basic engineering problems must be addressed:

1. Determination of the size of the system based on rational design criteria
2. Engineering of liners and earthen dikes
3. Distribution and collection of wastewater within the wetlands cells

The first problem is addressed in previous sections. This section is concerned with the implementation of the design and the problems that must be overcome in dealing with the second and third problems.

Structural Components

Constructed wetlands are very simple structures with only a few components requiring the attention of the designer. They are:

Liners and berms
Flow distribution structures
Level adjust structures
Distribution and collection piping

All of the structural components and materials have been used in wastewater stabilization or aerated lagoons. Frequently, the designer can provide solutions that are free of moving parts.

Typically, wetlands systems have flow monitoring requirements or perhaps a pump station for recirculation. The structures normally used for these function can be included in the design.

Subgrade Preparation

Preparation of the subgrade is a crucial part of the construction process. It is extremely important that the subgrade be properly compacted, and if the subgrade contains sharp rocks or rocks larger than 1¼ inches, geotextiles should be placed on the subgrade if liners are used. Larger rocks must be removed, or sand placed on the subgrade to prevent bridging of the liner.

Subgrades, when excavated from undisturbed soil, should be compacted to a minimum of 85 percent modified Proctor density. This will reduce settling under the liner and avoid the consequent stress. The total weight per square foot of material on the liner is approximately 450 pounds when the mass of the plants is included. Some soils will not support this weight without deformation; this should be considered when locating the wetland.

Some portion of the earthwork may be compacted fill because of site considerations. The level of compaction should be increased to a 90 percent Proctor density to minimize deformation.

It is essential to include some grading tolerance specifications. Before placement of liner materials, rough grades should be graded level or have a uniform gradient that is within ±0.1 foot. Heavy construction equipment must be kept off rough grades if the compacted subgrade is unable to support the vehicle's weight without deformation. Finish grades, gravel, or soil should be graded to within ±0.05 foot.

Berms and Liners

Berms and liners provide the basic containment structure of the reactor volume that is the constructed wetland. The structural and watertight integrity of the liner and surrounding berm is essential. Failure of either will result in loss of water, potential water pollution, and potential loss of plants as the water level declines.

Most of the difficulties encountered in the installation of liners in constructed wetlands occur because of the requirement to place soil and/or gravel on top of the liner without destroying the integrity of the liner. When choosing a liner, this problem should be considered carefully and preferably discussed with the contractor. Thicker is better; liners with embedded netting (scrims) are even better.

Berms Berms are a common element in wetlands construction. Generally, the height of the berm will not exceed 1–2 feet above the surrounding terrain on the upslope side. Upslope berms are used to divert surface runoff. Downslope berms are designed to retain the gravel bed and/or maintain the level of the wetland. Berm compaction should be 90 percent of the maximum Proctor density. Internal slopes of 1:1 should be considered a maximum if covered with liner material or erosion control blankets. External slopes should not exceed 3:1 without some erosion control. Slopes of 5:1 or less should be considered if muskrats are likely to be present. Biologists have discovered that the shallower slopes are not attractive to muskrat tunneling (staff biologist, Bosque del Apache National Wildlife Reserve, New Mexico.)

Occasionally, berms are constructed to prevent flood damage to the wetland. Berms then become flood control levees and should be designed accordingly, with rip-rap protection, vehicle access for inspection (10-foot width on top), and erosion control planting for berm stabilization. The minimum width on top should be 24 inches for foot traffic.

Liners The use of synthetic liners or clay has become a general requirement for the construction of wetlands in most states. Their usual purpose is to protect the groundwater and to ensure that the wastewater receives the required treatment before discharge into the groundwater, streams, or land application site.

Materials that are generally used include, but are not limited to, the following:

Polyvinyl chloride (PVC)
Polyethylene (PE)
Polypropylene (PPE)
Soil
Compacted clay and clay (bentonite) with scrim

These materials can include a scrim, which is a woven net of nylon or PPE embedded in the plastic material or enclosing the clay. Scrims provide extra strength and resistance to tears in the material. Liners with scrims will cost more. Liner materials with scrims are often marketed under trade names such as Hypalon or XR-5, or are trade products such as reinforced PPE.

The installation and testing of liners is an important element of a successful wetland project. Designers should become familiar with the installation process, the range of liner materials, and the different sets of specifications associated with each material. An excellent trade publication that should be required reading for designers using liners is *Geotechnical Fabrics Report 1996 Specifiers Guide.*

Table 4.1 rates the relative ease of installation for the various liner materials for large jobs (over 100,000 square feet). This scale is based on ratings provided by experienced installers (firms that have installed more than 2 million square feet), who have been trained in the installation of these liner materials, and who have all the necessary equipment.

Specifications should include a requirement for written approval of the subgrade by the liner installer before installation of the liner. Written acceptance of the subgrade preparation by experienced liner installers is usually a condition for maintaining manufacturers' warranties.

PVC Liners Each material has a unique set of properties that are worth consideration. For a given thickness, PVC is generally the least expensive material and the easiest to work with in the field. PVC materials to be used in constructed wetlands should be 30 mil or thicker. PVC has the best puncture resistance of all the materials (except those with scrims) and good friction properties. It is also available in large prefabricated pieces. It has the most flexibility but also has the least resistance to ultraviolet (UV) degradation. PVC must be covered to protect it from degradation. Since the wetlands construction process generally covers the liner with soil or gravel, PVC is a good first choice. Because PVC requires chlorine in the manufacturing process, it is therefore more environmentally unfriendly.

TABLE 4.1 Installation Difficulty (1 = Easiest)

PVC	PPE	PE
1	3	6

Polyethylene PE comes in two forms for liners: linear low-density PE (LLDPE) and high-density PE (HDPE). HDPE is harder to work with in the field than LLDPE, especially in cool weather. Both are more difficult to work with than PVC. The minimum thickness should be 40 mil (equivalent to 30 mil PVC for puncture resistance). UV resistance is good for both PE materials. Field repairs are easy. Seaming with tapes that will provide good waterproofing is easily performed in the field. The absence of chlorine in the manufacturing process makes this material more environmentally friendly than PVC. Its puncture resistance is not as good as that of PVC, and care must be exercised when placing the liner on soils with rocks, roots, or caleche. PE is usually 10 percent more expensive than PVC. Because of its use in landfills, regulators are more likely to be familiar with the properties of PE.

Polypropylene PPE's use as a liner is recent. As prices have dropped, its installation properties have encouraged the use of this material. Field seaming and field repairs are easy. PPE has good friction qualities. It has the best puncture resistance of all three plastics. PPE is available in large fabricated pieces, and it has excellent UV resistance. It is manufactured without chlorine and is probably the most environmentally friendly of the three materials. Its major disadvantage is that it is the most expensive of the three materials.

Liners with Scrims If the project can afford them, liners with scrims such as 60 mil Hypalon or 60 mil XR-5, or 45 mil PPE are clearly the toughest, and the most resistant to punctures, tears, and UV radiation. Vehicles such as front-end loaders and trucks can drive on them. This presents certain advantages in placing the gravel media (subsurface flow wetlands) or soil (surface flow wetlands). The difference in the cost of liner materials may be offset by the contractor's savings in placing the gravel media or soil.

Hypalon Hypalon has long been an industry standard. It is a proprietary product manufactured from chloro-sulfonated polyethylene with a nylon scrim. Hypalon has excellent UV properties, excellent puncture resistance, and good strength, and is available in large prefabricated sheets. But it is more expensive, and repair after aging is difficult. The loss of binder in the material makes repairs very difficult and expensive.

XR-5 XR-5 is a proprietary product. It is a PVC material with a scrim and has very high strength and excellent puncture resistance. It is available in large prefabricated sheets and is easy to repair. XR-5 is the most expensive liner and requires specialized equipment for installation. It is subject to some UV degradation and is made with chlorine.

Reinforced 45 mil PPE Reinforced 45 mil PPE has all the advantages of PPE but higher strength. It is UV resistant, has excellent puncture resistance, is easy to install and repair, and is manufactured in large sheets. It is not a proprietary product and is produced by at least three different manufacturers, so pricing is competitive. It is more expensive than 30 mil PVC.

Geotextiles Geotextiles are highly recommended if the subgrade cannot be prepared to the standards described previously. Geotextiles are easier to place than sand and have the advantage of staying on sloped berms. The following general design guidelines for the use of geotextiles should be considered:

1. If the visible rocks are less than ¾ inch, 4-ounce nonwoven polyester or PPE geotextile fabric should be placed on the subgrade to protect the liner from punctures.
2. If the rocks are larger than ¾ inch and smaller than 1¼ inches, an 8 oz geotextile should be used.
3. Rocks larger than 1¼ inches should be removed, or an underlayment of sand should be placed on the subgrade to prevent bridging of the liner material over the rocks.

Soil and Clay Liners Many of the first municipal constructed wetlands systems were installed without plastic liners. Later on, concerns about groundwater pollution imposed very expensive soil testing requirements on projects that do have soils suitable for compaction. Consequently, the costs associated with testing and soil compaction have driven the cost of using in situ soil up to the cost for PVC.

An alternative to plastic liners is the use of bentonite. Depending on soil characteristics, an amount of bentonite specified as pounds per acre is mixed with the existing soil, wetted, and compacted to make a liner whose characteristics will meet the percolation standard established by the regulatory agency. Costs depend on the in situ soil, and the distance from the bentonite supply. This material is mined, and deposits are limited to certain states, principally Wyoming, Montana, California, Arizona, and Colorado.

Finally, there are clay liners composed of a layer of clay between two scrims of finely woven PPE or PE. This liner material was developed for use in landfills where its chemical resistance properties are of primary importance. Costs in the West are similar to those of PPE. Shipping will add significantly to the cost as the distance increases from the Western sources increases. The puncture resistance of clay liners is generally low, and subgrade preparation is extremely important. This material is worth consideration in industrial applications or where landfill leachates are being treated by wetlands.

Inlet and Outlet Structures

Constructed wetlands require structures that can distribute wastewater uniformly into the wetlands, control the depth of water in the wetlands, and collect the treated effluent leaving the wetlands. Most structures are very simple, and several examples have been included for reference in Figures 4.4–4.10. In designing these structures, ease of construction, ease of maintenance, and operator safety and visibility should be the primary considerations. Other examples of these types of structures can be found in engineering textbooks on irrigation and in the U.S. Bureau of Reclamation's engineering standards.

Flow Distribution Structures

For gravity flow situations, simple V notch or horizontal weirs will work very well. These structures should be covered with lockable lids that are easy for the operator to open and inspect. Figure 4.4 is an example of such a structure. Construction is concrete block on a concrete slab. The use of precast units should also be considered. Often, precast units for underground construction will cost less than site-built units.

For small systems, there are several manufacturers of flow splitters such as that shown in Figure 4.5 that use reinforced PVC for the housing and PVC piping drilled with holes. These types of units work well with small flows (10,000 gallons per day or less). Because the PVC enclosure is fragile, it must be protected from accidental contact with mowers or other equipment.

Flow Distribution Piping

Once the influent has left the flow distribution structure, the wastewater must be uniformly distributed in the front end of the wetlands. For subsurface flow wetlands, distribution piping can take the form of perforated piping placed in the bottom of the wetlands. Alternatively, in warmer climates where freezing is not a problem, a series of adjustable tees placed on top of the gravel will distribute the wastewater equally.

In subsurface flow wetlands, other alternatives for the distribution piping are large leach field distribution chambers such as those manufactured by Infiltrator Systems, Inc., or 12- to 24-inch-diameter perforated PE drainage piping manufactured by Advanced Drainage Systems, Inc. Large-diameter, half-pipe sections have also been used successfully.

For surface flow wetlands, similar techniques will also work. The advantage of placing the distribution piping on the bottom is that there will be a uniform hydraulic head over each hole. Distribution can be improved by the installation of additional tees, as shown in Figure 4.6.

Flow Collection Piping and Level Adjust Structures

Flow collection is the reverse of flow distribution, and similar piping can be installed in reverse. Piping will enter level adjust structures as shown in Figures 4.7 and 4.8. The purpose of the level adjust structure is to allow the operator to adjust the water level to the desired height and to provide for a hydraulic gradient.

In a 1988 U.S. EPA manual on constructed wetlands, *Constructed Wetlands for Wastewater Treatment,* it was suggested that hydraulic performance could be improved by placing wetlands on a slope. In theory, this is a good idea; however, in practice, there are several problems. The first problem is that the idea is based on a uniform flow through the wetlands (surface and subsurface wetlands both have hydraulic resistance). During low-flow periods, such as occur during system start-up or during the daily cycles, the water in the front end of the wetlands will drain to the lower end faster than the incoming water can fill it. Once the plants have been established, this is not a problem, but during startup, this is a significant issue, because the roots of the new shoots can dry out.

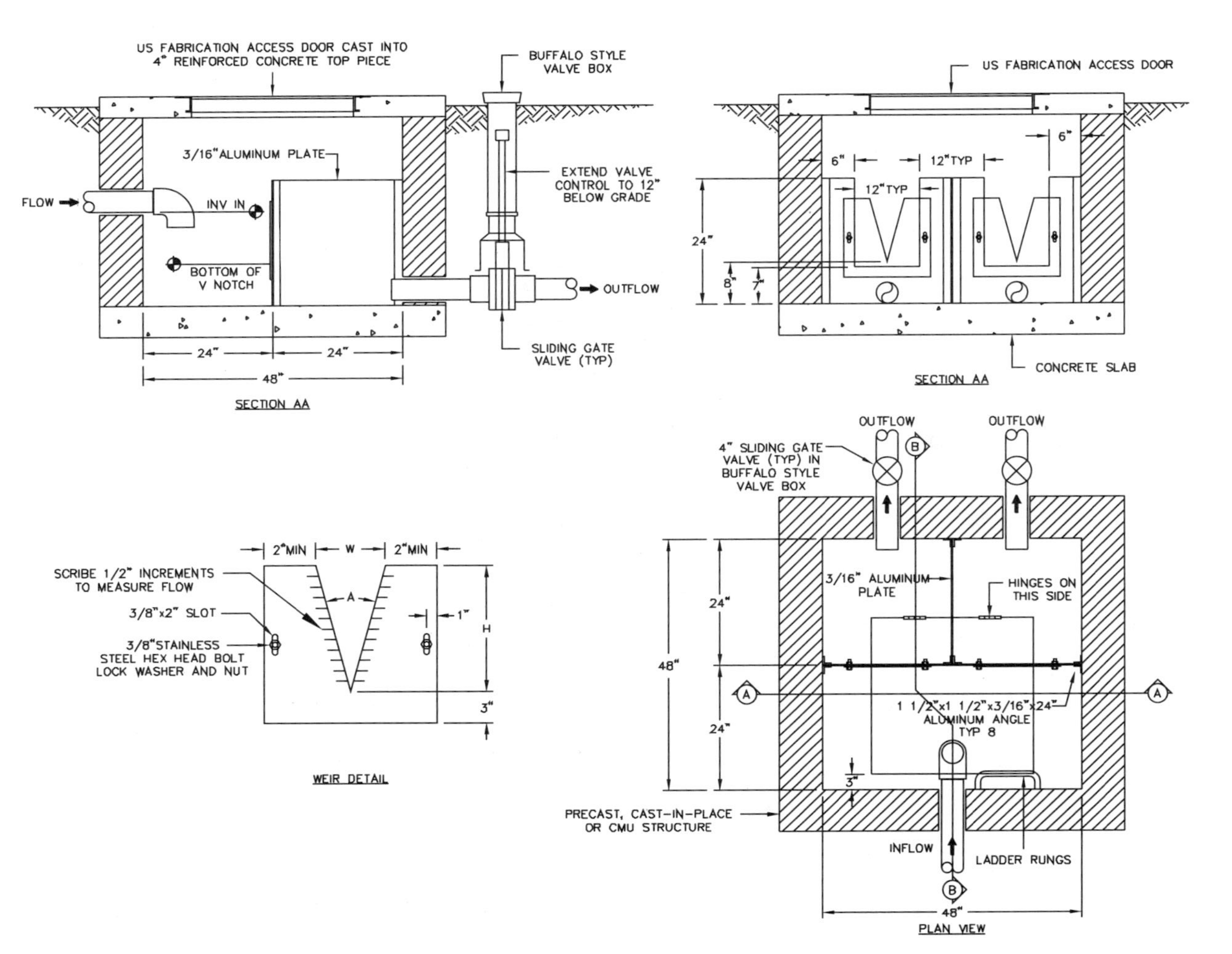

Figure 4.4 *Flow splitter/CMU (Courtesy of Southwest Wetlands Group).*

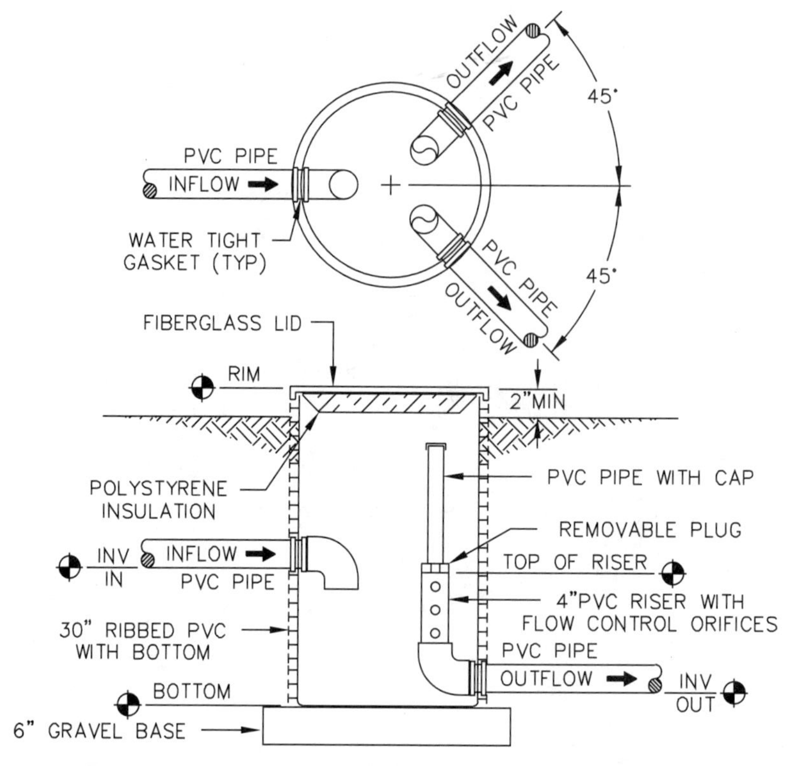

Figure 4.5 *Flow splitter/PVC (Courtesy of Orenco Systems, Inc.).*

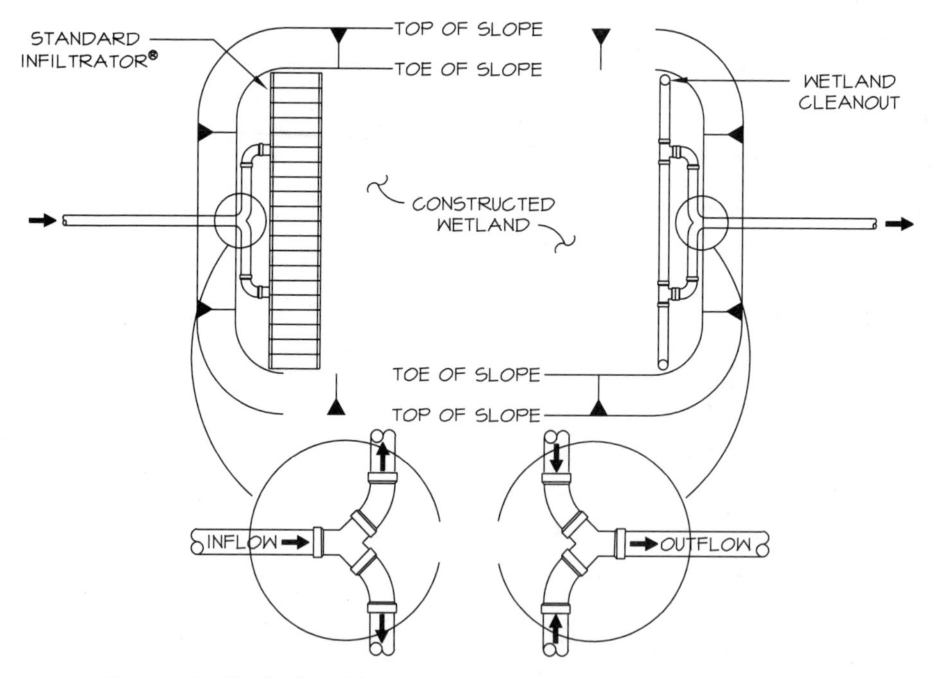

Figure 4.6 *Distribution piping (Courtesy of Southwest Wetlands Group, Inc.).*

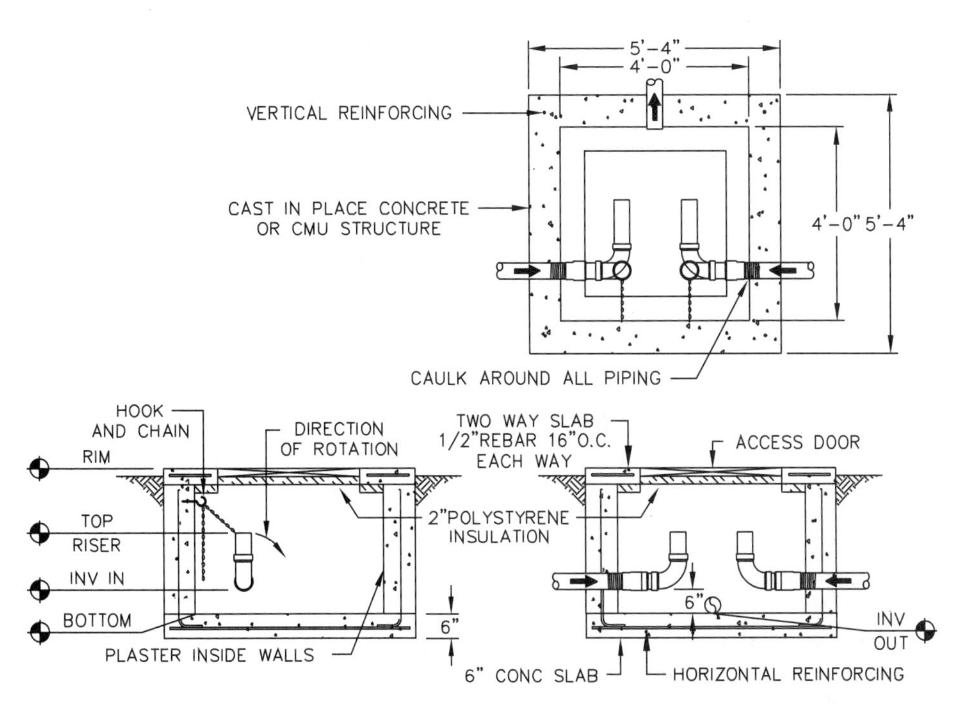

Figure 4.7 *Level adjust with swivel pipe (Courtesy of Southwest Wetlands Group).*

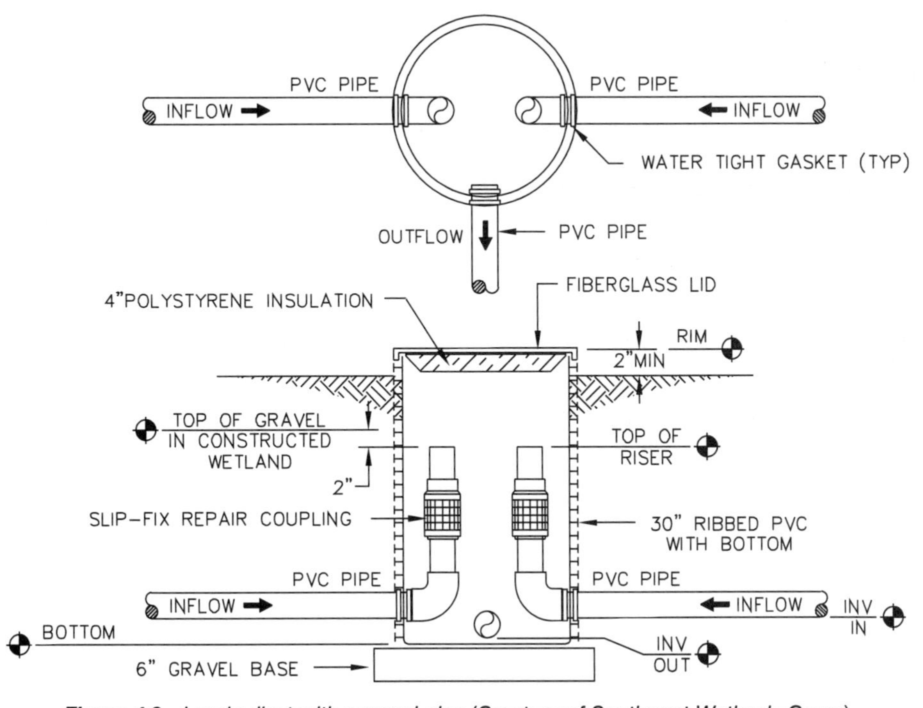

Figure 4.8 *Level adjust with removal pipe (Courtesy of Southwest Wetlands Group).*

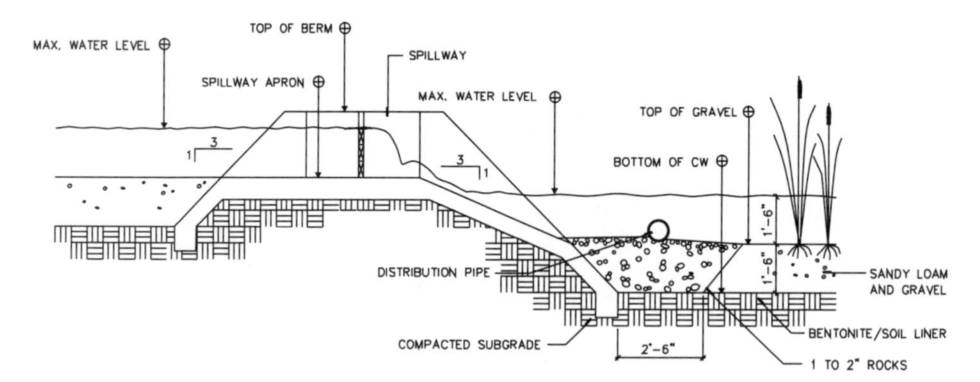

Figure 4.9 *Level adjust with cascade* (*Courtesy of Southwest Wetlands Group*).

If the bottom is sloped, then all the water drains to the low end and there is insufficient water for plant growth. For example, a 100-foot-long wetland with a 1 percent slope, will have water standing 24 inches at the back end and 12 inches at the front end.

In practice, it is much easier to set the hydraulic gradient by using adjustable pipes on swivels (Figure 4.7) or removable pipes of different length (Figure 4.8) that allow the level to be adjusted to match the flow. The pipes can be adjusted to match the changing hydraulic resistance as the wetlands matures.

When wetlands are placed on sloping ground, multiple cells are sometimes constructed to reduce the excavation and earthwork costs. Level control structures can be combined with cascades to add aeration capabilities. Figure 4.9 is an example of such a combined structure.

Other Structural Components

Debris Screens Surface flow wetlands typically drop a lot of leaves, and, during storm events, entire plants occasionally will be uprooted and float downstream to the collection piping or outfall structures. Unlike algae or duckweed, this larger

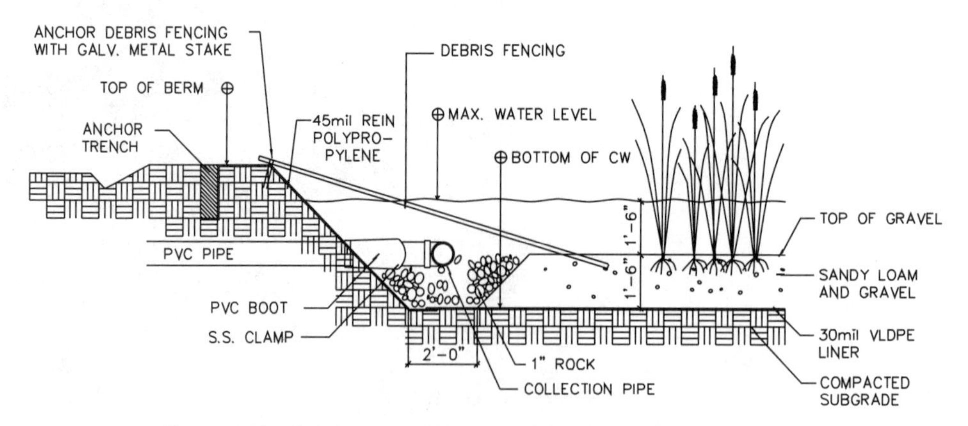

Figure 4.10 *Debris screen* (*Courtesy of Southwest Wetlands Group*).

debris will cause the collection piping to plug. Debris screens should be placed in front of these structures.

Establishment of Plants The three main genera of plants that are most commonly used in the United States are cattails (*Typha* spp.), reeds (*Phragmites* spp.), and bulrushes (*Schoenplectrus* spp. or, alternatively, *Scirpus* spp.). Many other species have been used, and no doubt additional ones will be proposed. Ideally, the hardstem species native to the region of the project should be used. Newly harvested plants from nursery stock, or from nearby drainage or irrigation ditches, will ensure that the plant genetics have become adapted to the local climate. Establishing new plants will require attention to the following:

Planting techniques
Season
Water depth
Viability of the plant

Ideally, planting should take place in the early spring, at least six weeks before wastewater is introduced into the wetland and after the last hard frost. In practice, planting has been completed the day before system startup and even, on occasion, after wastewater has been introduced into the wetlands. This latter situation should be avoided if possible, because it means that the planting crew will be placing plants in wastewater and will need to wear gloves and rubber boots.

The success of planting becomes more problematic as the season progresses, with late summer/early fall probably being the worst time. The reason for this is that the cuttings or transplants will be stressed from planting and will not have enough time to recover before the first frosts. Plants in this condition are in the worst possible condition for survival. It is better to wait until the plants become dormant and to plant after the first hard frosts. Dormant plants will have a greater chance of surviving throughout the winter.

If planting cannot be completed until late in the year, some nurserymen prefer to wait until the next spring. The rationale is that the plants are not providing treatment, so it is better to wait until the ideal planting time and improve the survival rate. As an interim measure, a green manure crop such as winter rye can be seeded in the gravel bed.

Gravel beds should be mulched to provide some insulation for the newly rooted plants and protection from the very high surface temperatures that occur on bare gravel surfaces. Water levels should be maintained at 1 or 2 inches below the surface in the gravel beds. In surface flow wetlands, the water level should be raised gradually to the operating depth, starting at 1 or 2 inches for new shoots. These small shoots can drown or be floated away by water depths that are too high or raised too quickly. If planting occurs in the spring, the plants will grow very rapidly during the summer, reaching heights of 6 to 10 feet in the first year. Some species of cattails (Typha domingensis) will grow to 15 or 16 feet, and giant cane (Arondo donax) will exceed 20 feet.

Typically, cattails, reeds and bulrushes are set out in the gravel bed of the subsurface flow wetland or the soil of the surface flow wetland as a rooted shoot approximately 12 to 18 inches tall. If these shoots have been grown in a nursery from cuttings or seeds, they will be delivered in bunches of 25 to 100 plants. They should be shipped in a refrigerated container or packed with ice and kept moist. Each shoot must have at least one viable growing tip visible. It is extremely important to keep the roots of these plants wet and to plant them in water as soon as possible. Wetlands plants are extremely susceptible to drying out, which will greatly diminish their chances of surviving. Contractors often are not aware that the normal habitat of wetlands plants is in saturated soils. They overlook the fact that temperatures on gravel beds can exceed 140°F. Specifications should direct the contractor to place the plants in buckets of water and to be pulled from the bunch as they are needed for planting.

For large projects, machines are available that can plant in soil or gravel at a rate exceeding 25,000 plants per day. Each plant will receive a dose of water at the time of planting. After a day's planting has been completed, the planted area should be flooded or the water raised in the gravel bed. Often after planting, water levels are neglected and the plants are left to dry out and die. Maintaining water levels around roots is extremely important to the survivability of the plants.

Gathering "free" plants from irrigation or drainage ditches is an attractive, low-cost means of obtaining plants. On occasion, mature plants have been harvested from drainage or irrigation ditches or from highway rights of way. If available, these clumps of mature plants must be separated and growing shoots selected for planting. These new shoots must be cut away from the mature mass. Each shoot, when cut out, must include a complete length of rhizome, that is, one joint that is similar to crabgrass, except that these rhizomes are much larger. A length of rhizome must include both shoot and roots. After separation, the cuttings must be kept wet and prevented from drying out before planting. Since these mature clumps of plants are extremely heavy, equipment such as front-end loaders or backhoes will be essential in the harvesting operation. Unfortunately, the apparent low cost of plant material will quickly be replaced by the reality of the labor intensity of this harvesting operation. Nursery plants will then appear to be a very attractive alternative.

Nursery plants are generally grown in saturated sand beds that allow a harvest of 12- to 18-inch shoots that can be bundled and packaged for ease of shipment and transplanting. These plants are shipped as bare root stock, and must be shipped chilled and kept wet. Most nurseries can meet these conditions. Some species, such as Phragmites, are grown and transplanted to gravel beds more successfully using peat pots. Peat pots are more expensive, but the survival rate in gravel beds is much higher. Transplanting is less successful in gravel than in saturated soils.

In terms of survivability after planting, cattails are the most hardy; bulrushes rank second and reeds, third. Reeds are more likely to survive if planted in muck or fine soils. Their fine root structure and lack of stored starch lead to higher mortality than cattails or bulrushes. The roots are extremely susceptible to drying out. However, once started, reeds often outperform both of the other plants and can crowd them out. Reeds grow well in gravel beds and poorly in surface wetlands. Cattails and bulrushes are ideal in surface flow wetlands.

Questions of plant diversity and appropriate species are worth considering. The proponents of reeds argue for a monoculture of reeds, while others argue that bulrushes are superior. Survivability of each species will depend on other factors, such as plant pests. Muskrats love cattails and bulrushes, while reeds apparently are inedible. On the principle that a monoculture is ecologically not recommended, a mixed planting of cattails, reeds, and bulrushes will at least let the process of natural selection decide the issue. Wetlands that are apparently cattail dominant can be succeeded by reed wetlands as climate, operations, and pests create conditions that are favorable to reeds. Darwin's principle of the survival of the fittest will determine the outcome, which can and will change over the life of the system.

Gravel, Sand, and Soil: Planting and Treatment Media

Subsurface Flow Wetlands With a little thought and reference to Graph 4.1, it is apparent that the best combination of hydraulic conductivity and void ratio can be found in the range of gravel between ½ inch and 1 inch in size. This is the preferred media size. It is relatively easy to obtain in most parts of the United States and is generally available by screening gravel used for concrete.

However, gravel is not the ideal material for the front end of the system because of the void *size*. Recall that most solids settle out in the front end of the wetlands, so larger void sizes provided by 1½ to 3-inch rock allow space for these solids to settle out and be digested without unduly affecting hydraulic conductivity. This rock, when used in the first 3 feet of a subsurface flow wetland, also promotes the uniform distribution of effluent. As effluent moves through the greater hydraulic conductivity of the larger rock and subsequently meets the lower conductivity of the ½- to 1-inch rock, this creates a hydraulic gradient similar to that of water running in a gutter that goes from steep to shallow. This "jump" places a uniform, albeit small, positive pressure against the entire face of the treatment media (½- to 1-inch gravel). This effect can also be accomplished by using proprietary products such as the Infiltrator^{tm} or ADS perforated PE piping, in lieu of large rock.

Selection of the medium is a critical factor in subsurface systems. Unfortunately, the ideal material is not always available, and compromises must be made. To reduce costs, locally available gravel materials should be used. Materials used in concrete are generally a good starting point, as they are usually available everywhere in the country. However, to ensure that the available materials will be suitable, a sieve analysis should be made and reviewed by the designer before specifying the material.

Although no hard rules can be given, the general principle is to select materials that are within a few sieve sizes. For example, specify gravel between ½ and 1-inch, or pea gravel from the #8 to the ⅜-inch sieve. This will produce material with the greatest void ratios; testing routinely shows void ratios greater than 40 percent. Other combinations can be developed by examining the gravel pit's standard sieving operations.

Gravel specification should have a maximum permissible percentage passing the #100 or #200 sieve. Either 1 or 2 percent is ideal, and washed, river-run gravel can often meet this specification. This specification can be relaxed if the 1 or 2 percent standard is difficult to meet. Excessive amounts of fines will settle out in the haul from the pit, and when dumped into the wetlands, these fines will drop out of the bottom of the truck and form small dams in the gravel bed. Gravel should also have a minimum Mohs hardness of 6. This will also decrease production of fines during transit and placement.

Pea gravel placed on top will ease the planting operations if larger rock is used as the medium. The same principle used for construction of sand filters applies when sizing materials. It is easier to plant in pea gravel than in 1.5- to 3-inch rocks. The use of larger rocks than those selected for the primary treatment media at the distribution and collection ends of the system will assist in the equal distribution of wastewater in the influent and effluent.

Placement of gravel over a liner is a challenge. If the gravel is crushed, the liner must be protected with a geotextile. In large projects, vehicles with large tires such as front-end loaders work well. The construction of a gravel road through the middle of the system that is at least 8 inches deep can provide a thoroughfare for bulldozers to push material. In any event, liner integrity must be maintained.

Gravel should be placed level, and should be free of vehicular tracks and depressions or ridges. The gravel bed can be filled with water to determine if the surface is level. Depressions can then be filled and ridges leveled.

Surface Flow Wetlands If the wetland is constructed on a compacted clay or plastic liner, some soil will have to be placed on the liner to provide for plant root development. If cattails are the primary plant, then 12 inches of a sand/clay/loam mixture will be ideal. Usually, soil found on site will be acceptable. This material should be free of organic debris and large rocks (>1.5 inches). It should be placed on the liner material so that the finished surface is level, free of vehicular tracks, and compacted to 85 percent of the maximum Procter density. Compaction of 85 percent can generally be accomplished by dragging the bucket of front-end loaders over each 6- to 8-inch lift of soil. If the soil is not sufficiently compacted, the cattails will float free. If it is compacted too tightly, then planting becomes difficult and plants will not root properly.

Bulrushes should have an additional 6 inches (18 inches total) for rooting. Once the soil is in place, the plants can be planted and water introduced. Water levels should be increased gradually to match plant growth. For very long cells, intermediate, temporary soil dikes should be constructed to maintain the water level as planting takes place. Planting should begin at the outlet and progress via these temporary dikes toward the front.

OPERATIONAL AND MAINTENANCE CONSIDERATIONS

Wetlands are easy to operate. They are essentially wastewater stabilization lagoons that have been planted or filled with gravel and planted. In most operational respects they are similar. Operators must be concerned with the same issues:

Maintenance of the water level
Maintenance of berms and embankments
Maintenance of watertight integrity
Control of nuisance pests
Maintenance of flow distribution and level control structures

Level adjust structures and flow-splitting structures are essential for the operations of wetlands. These structures must be kept free of debris; if weirs are present, they must be periodically brushed to remove bacterial growth that may affect distribution in low-flow situations. Regular visual inspection will determine if flows are equal; if problems develop, difference in flows will be noticeable.

Weeds in subsurface wetlands can be controlled with springtime flooding or manual weeding of the gravel bed. Certain weeds such as thistle, prickly lettuce, or salt cedar will compete aggressively with wetlands plants. This is not necessarily bad from the treatment point of view, but neighbors will object if the wetland is a source of noxious weeds. Weeds on berms and embankments should also be removed.

Berms should be inspected visually for leaks by examining the external slopes below the gravel or water surface. Excessive or unusually dark green vegetative growth is a good clue. Trees growing near berms may penetrate the liner; if they are growing on the eastern, southern, or western sides they shade the plants and suppress their growth.

Subsurface Flow Wetlands

It is extremely important to maintain the level in the wetlands as the plants grow. Once the plants have matured, lowering the water level to encourage root penetration is a possibility. This technique is practiced in Europe with the intention of encouraging root penetration as the roots are forced to seek the lowering water level. Other investigators (Gersberg, 1986; Reed, 1993) have found that roots will penetrate to their maximum depth, whatever the depth of water.

However, many operators have also found that draining the wetlands once a month during the growing season and then immediately raising the water level will draw oxygen into the wetlands. This is a beneficial process that assists in oxidizing carbon compounds, iron sulfides, and other anoxic compounds that have precipitated in the wetlands and that may suppress bacterial activity.

Harvesting of plants is not necessary; however, removing dead vegetation allows new growth in the spring to come in more vigorously. Leaving some leaf litter adds to the insulation value of the gravel surface. Harvesting at the peak of the growing season has been proposed as a means to remove nitrogen. However, the amount of nitrogen present in the leaf and stem mass at the time of harvesting does not justify the cost. It is better to encourage denitrification and let the bacteria accomplish this task. From an aesthetic point of view, annual harvesting of the plant litter in the fall will ensure vigorous and attractive plant growth in the spring.

A very strong argument has been made for the use of hard tissue plants such as cattails, reeds, bulrushes, giant cane, and bamboo because their litter does not leach the same amount of ammonia as soft tissue plants such as canna lilies. In the fall and spring, deamination from the leaves of soft tissue plants can add a significant ammonia load to the system discharge during these periods. Soft tissue plants (iris, canna lillies, daffodils, hyacinths, paper whites, etc.) tend to be those with attractive flowers or broad leaves that are desirable from a landscape architect's design perspective. These plants can be placed around the perimeter of the wetlands in a planting shelf (see Figure 4.3). In these locations, soft tissue plants will add to the attractiveness of wetlands without affecting performance. These plants require annual maintenance and removal of leaf litter to remain attractive.

Many operators of wetlands have noted that many birds, mammals, reptiles, and amphibians appear once the wetland has matured. Wetlands can be a significant food source for many animals. Wildlife can generally be regarded as a benefit, as the wildlife harvests some nutrients in the plants and subsequently distributes them over the landscape. However, some mammals, such as muskrats and nutria, can be a problem.

The worst offender is the muskrat. Muskrats like to make tunnels in berms and build nests in the middle of surface flow wetlands by harvesting plants. Tunnels in berms can be controlled by making the slopes of the berms 5:1 or less. Nests in middle of wetlands affect flow and may affect performance. Traps or local coyotes are the only means of control.

Insects can also be a problem. Grasshopper populations occasionally grow large enough to devour all of the cattails. Although this will not affect treatment, it will make the wetlands unattractive. Natural controls for reducing the number of grasshoppers such as sparrow hawks are ideal. Constructing perches for sparrow hawks will attract these predators, which in turn will help control grasshoppers. Bat houses and martin houses located nearby provide excellent control of mosquitos and other insects. These predators are responsible for removing significant quantities of nitrogen and phosphorus from wetlands, and this "free" removal system should be used by the operator (Mitch and Gosselink, 1992). Although wetland scientists have quantified this nutrient function in natural systems, it has not yet been done for artificial systems.

Mosquitos can be controlled in free water surface wetlands with the use of mosquito fish and other natural predators such as dragon fly larvae. Both predators are very successful and, in studies conducted at the Arcata, California, wetlands, mosquitos were less likely to be found in the wastewater treatment wetlands than in the surrounding mosquito abatement district (Gearhart, 1993). The design of free water surface wetlands is important. Open water areas to oxygenate the water are essential to support the invertebrate populations that prey on mosquito larvae. If the vegetation becomes too dense for predators to access the mosquito larvae, it should be thinned. Application of Bacillus thuringus or mosquito larvae growth attenuation hormones may also be necessary.

Surfacing of water in subsurface wetlands is a common problem that can be solved by several methods. Generally, the problem is the result of decreased hydraulic conductivity or settling of the gravel. In the first instance, hydraulic conductivity declines because of root growth and the accumulation of solids. If the system has been designed properly, the water level can be lowered a few inches. This will increase the hydraulic gradient, that is, increase the slope, and therefore make the water flow faster, thus overcoming the increase in resistance to flow. Completely draining the wetland will help oxidize solids and will also assist in improving hydraulic conductivity. If surfacing is the result of settling, this can be corrected by adding gravel to the settled areas.

The design of the system should include multiple cells, so that a cell can be taken out of service temporarily and allowed to rest if there are problems with surfacing. This can be done during the summer without affecting overall performance. In the worst case, the plants should be cut down and removed from the front one-third of the wetlands; in addition, the gravel and solids should be removed, and replaced with new gravel and replanted.

Surface Flow Wetlands

Water depth should be increased incrementally during startup to prevent young plants from drowning or floating out of the soil. Once the plants have become established, the operating level can be maintained throughout the year. It is essential that the maximum operating depth be maintained during the winter. It is highly unlikely that moving water at an 18-inch depth will freeze more than about 6 inches. Even in the extreme climates of the Colorado Rockies, the Wind River Range in Wyoming, and the Dakotas, surface flow wetlands have been operating successfully throughout the winter without freezing except for the top 6 inches.

The operator can consider dropping the water level in the spring to encourage the growth of new shoots. This will allow sunlight to penetrate more readily and thus signal the growth of these phototropic plants. As the new shoots clear the water surface, the operator should raise the water level. Dropping the water level will affect detention time and therefore may affect water quality. The increased growth may not be worth the effort. There are many cases in which the operator has done nothing and has obtained perfectly satisfactory plant growth every spring.

Wetland plants will reach an estimated maximum density of about 75 percent (density of plant material at the water surface). At this point, old shoots die out and new ones grow to maintain a steady state in population density. No plant removal is required. However, plant debris will float, clogging level control structures and overflowing weirs if not removed. This debris is especially likely to develop in the fall.

Duckweed and other aquatic plants will colonize the wetlands without operator assistance. These plants are beneficial and will help promote treatment. Algae are common at the beginning of operations, but as the plants reach maturity and duckweed colonizes the system, algal growth will be suppressed. Wildfowl and other animals will also show up and may interfere with operations.

Mosquitos can be controlled by seeding the wetlands with mosquito fish and dragonfly larvae. Placement of bat and purple martin houses will also add in the natural control of mosquitos and other insects.

Long-Term Operation and Maintenance Strategies

The major concern in operating and maintaining subsurface wetlands is the eventual clogging of the gravel medium. This is estimated from a very limited data set to be approximately 33 to 150 years. In surface flow wetlands, debris with a high carbon content, the precursor to peat, is assumed to accumulate at the rate of one millimeter per year. Therefore, berm height determines the useful life of surface flow wetlands.

It has been stated that refractory organic compounds with a high carbon content are removed by the plants to become leaves and stems, and are subsequently deposited on the surface of the gravel. In any case, no accurate predications about the useful life are possible. Clearly, communities will need some form of planning horizon; for the present, thirty-three years seems the best estimate that can be offered for subsurface flow wetlands.

Future developments in plant breeding are likely to provide F-1 hybrids with superior characteristics in the same manner that plant breeders have improved the strains of corn. Equally likely is the development of the use of cultured bacteria and micronutrient additions.

The ability to provide water with all the essential nutrients for aggressive plant growth should be considered an asset. The EPA has long supported land application of wastewater to grow crops and irrigate tree farms (U.S. EPA 1981). Wetlands can be considered a specialized form of land application with different species of plants for consideration. Designs that include treatment and resource production, habitat development, recreation, and, as in China and Java, food production will become more common. The U.S. EPA has provided an excellent publication that summarizes the approaches taken by seventeen municipal systems that have recognized some potentials of this technology (U.S. EPA 1993).

5

Stormwater Renovation with Constructed Wetlands

And Noah said to his wife when he sat down to dine, "I don't care where the water goes if it doesn't get into the wine."

—G. K. Chesterson, (*Wine and Water,* 1911)

There is a prevailing public attitude of general unconcern over what happens to stormwater, as well as wastewater, as long as it is out of sight. The preoccupation with the mere collection and conveyance of stormwater management has resulted in a paucity of creative multipurpose approaches to on-site stormwater management.

Stormwater management and treatment have very different design and management problems from those associated with treatment of wastewater. Instead of a reasonably reliable flow of water that is picked up at the source and conveyed in underground pipes to a treatment facility, stormwater may be conveyed on the surface or in pipes; has unpredictable flows within a wide range of potential storm events; and is more prone to conveying heavy metals, hydrocarbons, and other contaminants from parking areas and highways that may be absent from many wastewater treatment systems. The EPA's National Urban Runoff Program (NURP) established that in many cases the first flush of stormwater in an urban area may have a level of contamination much higher than that normally present in sewage wastewater. The reasons for this are due primarily to the tendency of the initial stormwater flows to pick up and transport much of the deposed vehicular, animal, and human detritus from pavement. Such materials include hydrocarbons and asbestos from brake linings, heavy metals such as lead and zinc, bird and animal wastes, herbicides and pesticides, and others.

ON-SITE STORMWATER MANAGEMENT

It is fair to state that most engineers and public officials view issues of stormwater management within a strictly utilitarian context. Stormwater has generally been handled like wastewater, in underground conveyance systems, and simply discharged to the nearest stream or lake rather than treated or allowed to infiltrate on site. The extent to which stormwater can not only be managed on site and treated by wetlands, but also provide attractive and educational features in the landscape, is rarely recognized, and only in a few relatively restricted regions of the country, at least at this time. In 1977, one of the first demonstrations of the effectiveness of a constructed wetland in the treatment of stormwater runoff was described in an EPA publication (Hickok, et al. 1977) dealing with urban runoff treatment methods. The report described the beneficial effects of aquatic plantings in reducing the pollutant load in a discharge of urban stormwater runoff into a lake in Minnesota. Further research on the Wayzata wetland provided the basis for the design of stormwater wetland treatment systems for a shopping center and an airport in the area.

Since then, there have been ample opportunities in other parts of the country to witness the effectiveness of "wet ponds" and constructed wetlands in filtering and removing pollutants from stormwater runoff. Hundreds of wetlands and ponds, with several variants, have been constructed for attenuating the stormwater pollution problem, particularly in Florida and the mid-Atlantic states.

Over the past decade or so, much greater interest in applying urban stormwater best management practices (BMPs) to runoff problems has developed. Although the range of approaches to BMPs is quite wide, depending upon specific circumstances, the use of presettlement basins, infiltration/filtration basins, dry ponds, temporary storage vaults or pipes, and many other techniques works quite well in controlling runoff and removing heavier suspended particles. However, the pollutants typically adsorbed or associated with finer particles such as nutrients, heavy metals, and petroleum hydrocarbons remain suspended and therefore pass on downstream into receiving waters. To effectively trap and remove these pollutants from runoff, other measures such as biofiltration (vegetated) swales, wet ponds, and constructed stormwater wetlands should be considered. Wetlands have shown removal efficiencies of 41 to 73 percent of lead, zinc, and TSS from urban runoff (Martin, 1988).

Section 402(p) of the 1987 federal Clean Water Act (CWA) required the U.S. EPA to establish National Pollution Discharge Elimination System (NPDES) stormwater permits, which created much stricter control of stormwater runoff than existed in the past. This "final stormwater rule"—which was not wholeheartedly embraced at an individual state regulatory level due to lack of personnel to investigate and enforce the law—was phased in between 1989 and 1994 in most states. It requires a stormwater discharge permit from municipalities and an erosion control plan for all construction development that affects areas of 5 acres or more.

This is distinct from the permits that authorize activities that occur in wetlands or waters of the United States, as required under Section 404 of the CWA and administered by the U.S. Army Corps of Engineers. The regulations affect these permits that are provided as nationwide permitting (NP) and regional general permitting (RGP), along with individual permits for projects that cannot be approved under NPs or RGPs. Although mitigation may not be required under the NP or RGP permits, it is required under all individual permits. The regulations guiding the application of various Section 404 permits and the classification of activities have undergone considerable revision, which is continuing at the time of this writing.

It is widely recognized that the key to controlling damage and pollution from stormwater runoff is management and treatment, insofar as that is possible, at the source. For nonpoint source situations, such as runoff from a building or set of buildings, parking areas, and so on, many municipalities have responded to the tighter EPA regulations by enacting ordinances intended to address point source runoff. Unfortunately, these well-meaning efforts have focused primarily on detaining runoff at the site, untreated, rather on infiltration or measures to improve the quality of the water through natural means of treatment. Most major cities in the West now have ordinances requiring detention of all increased runoff created by new construction. Almost without exception, however, the result of these types of ordinances has been the creation of unsightly depressions in the landscape, often lined with cobble rock and usually presenting a significant weed problem. An extreme example of the potential for unfortunate aesthetic results of such ordinances is to be found in at least one major city in the Southwest, Albuquerque, New Mexico, where any detention pond designed to store water temporarily to a depth exceeding 18 inches must be entirely fenced with a barrier at least 4 feet high, which inevitably ends up creating a hole in the ground surrounded by a chain link fence.

In the Denver metropolitan area, detention basins typically fill in with cattails on their own. However, there are some examples of well-designed basins skillfully integrated with the overall landscape, with a diversity of plants including sedges and grasses in the transition zone between the wetland and upland landscape (Figure 5.1).

It is axiomatic that innovations are rare unless some entity takes the lead and assumes the risk of failure by attempting new solutions to old problems. The risks in developing untested design solutions are very real, and all professionals worry about their potential liability should an innovative design fail. Not only is the

Figure 5.1 *Stormwater detention basin, Aurora, Colorado (Craig Campbell).*

United States the most litigious society in the world, but professional liability insurance underwriters are making sure that their insured firms are aware of every conceivable exposure for which they might be held legally responsible. This is a sound educational policy, but one that can create a bit of paranoia and certainly an overly conservative approach to projects that demand the application of innovation, imagination, and experimentation.

KEYSTONE AND COPPER MOUNTAIN RESORTS, COLORADO

One of the more fascinating examples of the changing public attitude toward wetland values concerns the significant role that wetlands can play, both functionally and aesthetically, in a resort setting. Several very interesting examples are to be seen at two ski resorts in Colorado.

At Keystone, there are examples of stormwater treatment wetlands; mitigation or replacement wetlands; an educational wetland park; and natural wetlands located primarily along the Snake River, which flows through the Keystone base village area.

With a long history of respect for both the habitat values and the visual qualities of the historic wetlands in the area, Keystone has integrated over 100 acres of natural and constructed wetlands into the fabric of the riverine corridor with boardwalks, interpretive signage by elementary school children, and other features celebrating the role of wetlands in the mountain environment (Figure 5.2). In one of the first pilot projects of its kind at a ski resort, Keystone/Intrawest contracted with Colorado's Department of Health to develop the Snake River Urban Runoff/Wetlands Park project as a high priority for nonpoint pollution control to reduce nu-

Figure 5.2 *Frostfire Wetlands, interpretive signage by elementary school, Keystone Resort, Colorado (Craig Campbell).*

trients and metals from urban runoff and to educate the public about the value of wetlands. Part of the rationale for the project was the need to remove phosphorus, metals, and other pollutants that the Snake River conveys into a very important fishery and drinking water source—Lake Dillon.

Now called the Frostfire Wetlands, this stormwater runoff treatment system incorporated first-stage sedimentation ponds, along with a newly created 1.5-acre (3.7 ha) constructed wetland and enhancement of existing wetlands to improve water quality and wildlife habitat (Figure 5.3). The system was designed to receive and treat base runoff from a newly developed base area and adjacent parking lots. Water quality monitoring and data collection have been done by Dr. William M. Lewis, Jr., of the University of Colorado Center for Limnology, and after two full seasons the results have shown significant removal of phosphorus. It appears that the wetland treatment is providing excellent results, with over 94 percent of phosphorus retention even during storm events. The performance of the system will continue to be monitored.

A key component of this project was the development of an education program sponsored by Keystone/Intrawest designed to teach students and citizens about wetland and riparian functions and their role in maintaining water quality and wildlife habitat. The education program includes the development of trails with interpretive signs and a teacher education program through the local Keystone Science School located nearby.

At Copper Mountain, also owned by Intrawest of Vancouver, British Columbia, the last significant parcel of land for residential development has been carefully inventoried to delineate approximately one-third of the property that is composed of wetlands. These are primarily seep wetlands dominated by low willow shrubs that occur on relatively steep slopes. Recognizing the attractiveness of these wetlands both as natural buffers and as wildlife habitat, Design Studios West worked

Figure 5.3 *Frostfire Wetlands Keystone Resort, Colorado.*

closely with Intrawest to develop a plan based upon a minimal-impact, low-density residential lot configuration attractive both to the "ski in, ski out" adherents and to those who seek relative isolation and connection to the natural environment. Included in the plan are a natural wetlands park, interpretive trails, and other features somewhat unusual within a ski resort setting.

STORMWATER MANAGEMENT IN FLORIDA AND MARYLAND

Three areas of the country that have supported the concepts of biofiltration and stormwater wetlands are the states of Florida and Maryland and King County, Washington. Florida has a fairly long history of using both natural and constructed wetlands for wastewater treatment, but the state also began to utilize them for stormwater management and treatment after the implementation of the Florida Stormwater Rule in 1982. The regulations implemented by the Florida Department of Environmental Regulation required all newly constructed stormwater discharges to use appropriate BMPs to treat the first flush of runoff; vegetated systems and wetlands were considered appropriate BMPs. The most common "wet detention" system design in Florida is a permanent stormwater pond, a temporary storage area above the permanent pond, and a littoral zone planted with native aquatic plants.

The Florida Department of Environmental Regulation published *Stormwater Management—A Guide for Floridians,* probably the best laymen's guide to problems of understanding and managing stormwater ever made available to the general public anywhere in the country. Covering in clear and reasonable detail the concepts of watersheds, the hydrologic cycle, and groundwater systems related to lake, river, and estuarine systems, the guide goes on to address effects of urbanization on stormwater quality, along with stormwater management practices that stress infiltration and the importance of vegetation, grass swales, infiltration trenches, wetland stormwater systems, and other management techniques. The BMP treatment "train" is discussed, along with federal, state, and local governmental stormwater permits (Figure 5.4).

Included in the chapter entitled "Principles of Stormwater Management" is a statement perfectly reflecting the approach taken in this chapter:

> Optimum design of the stormwater management system should mimic (and use) the features and functions of the natural stormwater system which is largely capital, energy, and maintenance cost free. Most sites contain natural features which contribute to the management of stormwater under the existing conditions. Depending upon the site, features such as natural drainageways, depressions, wetlands, floodplains, highly permeable soils, and vegetation provide natural infiltration, help control the velocity of runoff, extend the time of concentration, filter sediments and other pollutants, and recycle nutrients. Each development plan should carefully map and identify the existing natural system. "Natural" engineering techniques should be used as much as possible to preserve and enhance the natural features and processes of a site to maximize the economic and environmental benefits. Natural engineering is particularly effective when combined with open spaces and recreational use of the site, or in developments that use cluster techniques. Design should seek to improve the effectiveness of natural systems, rather than to negate, replace, or ignore them. (p. 29).

Figure 5.4 *Grass bifiltration swale, Florida (Florida Department of Environmental Regulation).*

Stormwater Management stresses the importance of on-site storage and infiltration through the integration of ponds and lakes, which "provides storage, environmental protection and enhancement of community amenities" (p. 30). Prior to discharging stormwater to surface or ground waters, the manual suggests that "Runoff should be routed over a longer distance, through grassed swales, wetlands, vegetated buffers and other areas designed to increase overland "sheet" flow. These systems increase infiltration and evaporation, allow suspended solids to settle, and help remove pollutants *before* they are introduced to Florida's waters" (p. 30).

Eric Livingston, Environmental Administrator, and Ellen McCarron, Environmental Specialist for the Tallahassee office of the Florida Department of Environmental Regulation, were responsible for preparing this remarkably progressive and readable manual. There is no question that the particular growth pressures, groundwater problems, and other local factors in Florida have stimulated the evolution of that state's regulatory framework. In 1989 Florida adopted stormwater legislation creating a statewide watershed management framework that has resulted in the establishment of local Water Management Districts and a comprehensive program operating on a cooperative basis with both local and state governmental agencies. Watershed Management Plans are expected to address the reduction of pollutant loads in stormwater systems, among other things, and the plans are to be implemented through local land development regulations addressing nonstructural preventive controls, septic tank siting and maintenance, buffer zones, and even nutrient and pesticide management. Public education is also stressed, such as conveying information to homeowners that fertilizing lawns followed by heavy watering causes nutrient problems in local waters. There is also a volunteer program of stenciling storm drain covers with signs stating "DUMP NO WASTES—DRAINS TO LAKE (RIVER, ESTUARY." Boulder, Colorado, is another city with educational and interpretive signage along the Boulder Creek trail system to explain the nature of stream ecology and the consequences of dumping contaminants into storm drains (Figure 5.5)

Figure 5.5 *Educational signage, Boulder Creek, Boulder, Colorado (Craig Campbell).*

WATERSHEDS AS PLANNING UNITS

Watersheds have recently become widely respected throughout the country as one of the most useful landscape and environmental units central to the documentation of natural resources for planning, zoning, and assessment purposes. Watershed management units are classified from the smallest and most localized—the catchment—through progressively larger categories of subwatershed, watershed, subbasin, and finally the basin. The quality and quantity of runoff from a catchment are almost entirely influenced by the nature of the development within it, and are therefore the primary focus for planning and design of the more innovative treatment concepts. The management focus for each unit varies, along with the responsible governmental planning authorities.

National support for the use of a watershed approach to site planning is demonstrated by the cooperation of almost 100 cities and counties in all regions of the United States in developing a volume (Schueler, 1995) addressing the subject, with a focus on stream protection. This report provides an excellent overview of the design of biofiltration swales and filter strips, porous pavement, new street designs that incorporate surface treatment of stormwater, buffers and ponds, and other topics.

An interesting and ambitious example of a watershed-based assessment of the feasibility of constructed wetlands is a report prepared by The Pineywoods Resource Conservation and Development, Inc., of Nacogdoches, Texas, a nonprofit, grass-roots organization serving a ten-county area, and the Trinity River Authority (Pineywoods, 1995). The report addresses the unique character and ecological parameters of the Trinity River watershed. It encompasses four distinct ecological zones and represents one of the first efforts to define in detail the specific local environmental factors that affect the design and operation of constructed wetland systems by utilizing all available data, such as those on soils, precipitation, and slopes, and by developing a composite map of areas where costs are lowest.

Models for constructed wetlands were developed for each of the four major *eco-areas,* with designs for both a typical single-family residence and one for a trailer park as illustrations. As low cost and sustainability were among the objectives of the study, areas with little slope that do not allow for gravity flow systems were generally ranked as less suitable for constructed wetland systems.

This admirable cooperative effort involving government agencies, nonprofits, and interested individuals was intended to provide designers and managers with a "unified language, procedure, and a set of tools to determine the feasibility of low-cost constructed wetlands in any watershed." What must be understood, however, is that the model for the Trinity River Authority is represented by the methodology of collecting all available data for this particular watershed. Every other area has a different set of environmental characteristics that need to be documented.

Going beyond even the potential for creating functional stormwater treatment wetlands that are also wildlife habitats, artists have entered the arena and broadened the inherent possibilities by incorporating stormwater treatment wetlands into public artworks that are envisioned as providing public interaction, education, and amenities by symbolically incorporating elements that reveal the underlying wetland biological processes to visitors. One of the most outstanding collaborative projects in this category, which involved an artist, landscape architects, and engineers, is the Waterworks Garden project at a water reclamation plant in Renton, Washington, which is described in detail in Chapter 9. Other art projects involving wetlands and wildlife habitat also described in that chapter mark a new era in incorporating the unique talents and perspectives of artists into public infrastructure projects that formerly were seen as best left "out of sight, out of mind."

It is only fair to point out that some landscape architects are also responsible for creating stormwater management features that can also be considered artworks. One project that clearly falls into that category is William Wenk's stormwater management project on Shop Creek in Denver, which is both an innovative stormwater management concept and an attractive sculptural art form in the urban landscape. This project is described later in this chapter.

One of the more significant examples of the use of constructed wetlands for stormwater management in Florida is a regional system created in Tallahassee in 1983 to reduce pollutant loads to Megginnes Arm and Lake Jackson. A watershed of about 900 ha that consisted of a mixture of residential and commercial uses is treated with a 28.4 ac. detention pond, a 4.7 ac. intermittent sand filter, and a 6.2 ac. constructed wetland marsh divided into three cells. Although the system performs well during the growing season, it is a net exporter of nutrients in the winter from dieback and decay.

A more recent example in Florida of the use of constructed wetlands for stormwater renovation is the Greenwood Urban Wetland in Orlando, a multiple-use facility designed from the beginning to provide an attractive park, walkways, and wildlife habitat in addition to stormwater treatment. A detailed description of this interesting project is provided later in this chapter.

WASHINGTON STATE STORMWATER GUIDELINES

In addition to the benefits provided by stormwater treatment wetlands, it has been proven that considerable pollutant removal occurs within grassy swales and biofiltration swales planted with a mixture of vegetation, often woody plants. Prof. Richard Horner of the University of Washington was one of the first researchers to demonstrate clearly the ability of grass swales to trap and remove a variety of pollutants, including heavy metals such as lead, from stormwater runoff (Horner, et al. 1982). The State of Washington Department of Ecology has published guidelines on the design of biofiltration swales in their *Stormwater Management Manual* for the Puget Sound Basin, with an emphasis on the use of infiltration measures and constructed wetlands as BMP detention systems. Biofiltration is considered to be a stand-alone treatment for conventional pollutants but not for nutrient control. The standard event for biofiltration design is the six-month event, consistent with the other BMPs.

The manual states that detention BMPs utilizing a permanent pool of water, such as "wet" ponds, vaults, or tanks, are considered the most effective treatment and notes that shallow marsh areas within a permanent pool provide additional pollutant removal. There is a caveat, however: the fact that we are still in an early stage of development with constructed wetlands for stormwater treatment that requires considerable refinement based upon regional climatic patterns. The manual states:

> The effectiveness of "shallow marsh" detention BMPs at treating nutrients is under investigation. Because the majority of rainfall in the Northwest occurs in the winter months, when biological activity is low, wet pond BMPs which utilize biological removal mechanisms may not function as effectively as in other regions of the country. The need to control nutrients and other "non-conventional" pollutants may require changes in the current BMP selection strategy and design criteria. Many of the design details given . . . have been taken from work on the East Coast and elsewhere and will be subject to refinement after further experience is gained in this region. Nevertheless, the details presented in this manual represent the currently available information and should provide a sound basis for design of detention BMPs. (p. III-4-4)

It is interesting to note that the manual makes a clear point regarding the environmental and visual effect of wet ponds or constructed wetlands within the overall landscape, a consideration otherwise totally lacking in most jurisdictions. Under "Planning Considerations," the Manual states:

> Wet ponds require careful planning in order to function correctly. Throughout the design process the designer should be committed to considering the potential impacts of the completed facility. Such impacts can be positive or negative and can be as broadly classified as social, economic, political, and environmental. Designers can often influence the positive or negative aspects of these impacts by their careful evaluation of decisions made in the design process. . . . If the facility is planned as an artificial lake to enhance property values and promote the aesthetic value of the land, pretreatment in the form of landscape retention areas or perimeter swales should be incorporated into the stormwater management facility. If possible, catchbasins should be located in grassed areas. By incorporating this "treatment train" concept into the overall collection and conveyance system, the engineer can prolong the utility of these permanently wet installations and improve their appearance. Any amount of runoff waters, regardless [of] how small, that is filtered or percolated along its way to the final detention area can remove oil and grease, metals, and sediment. In addition, this will reduce the annual nutrient load to prevent the wet detention lake from eutrophying. Detention system site selection should consider both the natural topography of the area

and property boundaries. Aesthetic and water quality considerations may also dictate locations. For example, ponds with wetland vegetation are more aesthetically pleasing than ponds without vegetation. Ponds containing wetland vegetation also provide better conditions for pollutant capture and treatment. A storage facility is an integral part of the environment and therefore should serve as an aesthetic improvement to the area if possible. Use of good landscaping principles is encouraged. The planting and preservation of desirable trees and other vegetation should be an integral part of the storage facility design. (p. III-4-21)

This, of course, is one of the most progressive approaches to stormwater management anywhere in the country, along with that demonstrated in Florida, Maryland, and King County, Washington. The manual does recognize, however, the significant differences represented by ponds and marshes within established urban areas, and offers some caveats under the heading "Nuisance Conditions" that are summarized in Chapter 7.

Under the heading of "Multiple Uses," which is an overall theme of this volume, the manual states: "Multi-purpose use of the facility and aesthetic enhancement of the general area should also be major considerations" (p. III-4-21).

KING COUNTY METRO

The Seattle (King County) Metro transit division is to be commended for their farsighted concern about the effects of urban pollution caused by the by-products of motor vehicle operation. These include petroleum hydrocarbons from lubricants, solvents, and exhaust; surfactants from cleaning fluids; antifreeze; and heavy metals from fuel additives, body corrosion, tire and brake wear, and so on, many of which adsorb to surface dust and flush with other surface residue into drainageways, streams, and lakes.

Among the few studies that have been undertaken specifically to determine the efficiency of particular wetland plants in assimilating stormwater pollutants is the *South Base Pond Report—The Response of Wetland Plants to Stormwater Runoff from a Transit Base* by Seattle's Metro agency. Recognizing that the pollutant removal efficiency of stormwater detention ponds can be significantly enhanced by converting them to constructed wetlands, the study set out to examine the potential for five species of wetland plants to absorb and accumulate hydrocarbons, lead, zinc, and total petroleum hydrocarbon (TPH). The plant species were *Typha latifolia, Scirpus acutus, Iris pseudocorus, Eleocharis ovata, and Sparganium spp.,* all of which displayed some capacity to absorb lead, zinc, and TPH. Interestingly, a nonnative ornamental wetland plant, the yellow iris (*Iris pseudocorus*), stored mean levels of petroleum hydrocarbons in their root tissues that were significantly higher (1,566 mg/kg) than those in surrounding mean soil levels (652 mg/kg), and the common cattail (*Typha latifolia*) accumulated the highest concentrations of lead (0.049 g/m^2), zinc (0.467 g/m^2), and TPH (9.96 g/m^2) per unit area of plant coverage. *Sparganium* spp. had the highest mean concentrations of lead and zinc per unit weight of tissue on a dry basis. Biomass measurements taken at the end of the growing season for each of the plant species indicated, as might be expected from experience elsewhere, that *Typha latifolia* produced the largest biomass per unit area (8.02 kg/m^2) of the five species, followed by *Iris pseudocorus* (3.97 kg/m^2), *Scirpus acutus* (3.13 kg/m^2), *Eleocharis ovata* (0.98 kg/m^2), and *Sparganium spp.* (0.63 kg/m^2). These measurements were undertaken to estimate the pollutant uptake potential per unit area of a single species. This experience from the Puget Sound area of Washington State is not necessarily applicable to the performance of any of these species in other soil conditions, loading rates, or climates, and at least some tests in the Midwest indicated that *Sparganium spp.* established more rapidly in Michigan than did *Typha*. A number of species were also tested to determine their response to a range of inundation levels, as plantings in a stormwater wetland must be selected both for their ability to absorb and concentrate pollutants and also to withstand considerable variation in levels of inundation or soil moisture. The summary recommendations of the South Pond Project are as follows:

Create low-sloping banks
Regulate water levels
Plant *Typha latifolia* near inflows
Manage the growth of *Typha latifolia*
Plant a diversity of species
Install other pollutant control devices
Plant perennials and plants with long growing seasons
Create a deep forebay

Included in the report is a wonderful graphic image (Figure 5.6) indicating the suggested placement of wetland emergent plant species in varying water depths, as appropriate for each plant species.

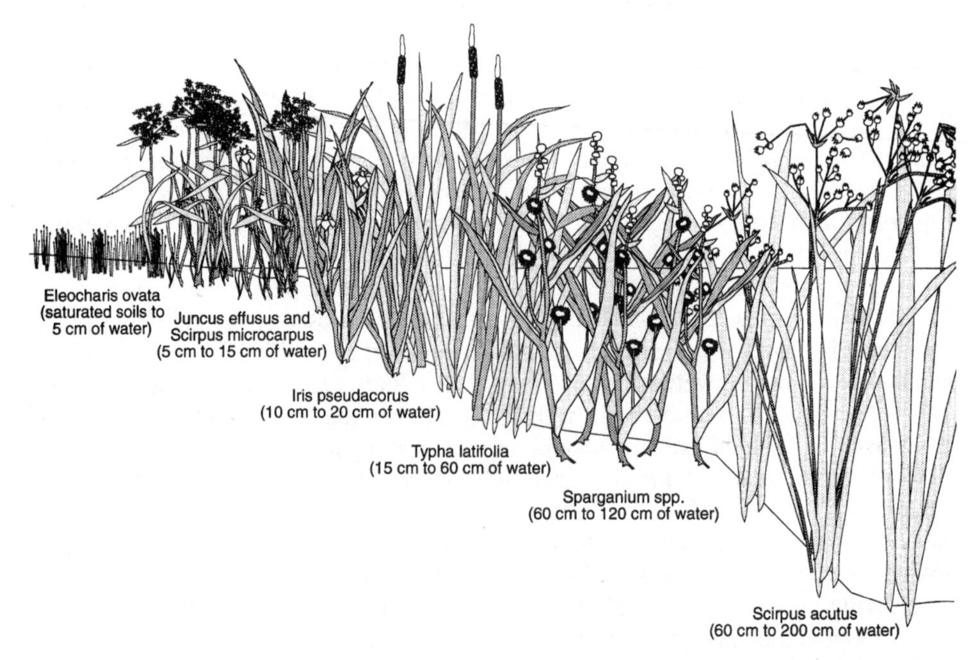

Figure 5.6 *Recommended stormwater treatment pond edge plantings (King County Metro).*

In Canada, studies in the city of Winnipeg, Manitoba, demonstrated that permanent impoundments with a minimum surface area of 5 acres were the most optimum stormwater control measure for the city. It was shown that the large ponds could control stormwater, reduce downstream pollution, and provide wildlife, recreational, and aesthetic values. At two facilities, one with eight interconnected impoundments, BOD reductions of 30–75 percent and suspended solids reductions of 85–94 percent were obtained. In addition, the studies stated that vegetation should be allowed to develop on a portion of each impoundment for both wildlife habitat and aesthetic reasons.

SEDIMENT INPUT

One of the characteristics of stormwater as opposed to wastewater is the sediment load often carried into any detention or retention areas. Clay particles contain negative electrical charges on their surfaces that tend to keep the particles dispersed and in suspension. Some aquatic plants have positive charges on their surfaces that tend to attract the oppositely charged clay particles and cause them to drop out of suspension. Although this process tends to clear up the water, there is a corresponding deposition that alters the pond environment as it becomes more shallow. Sedimentation basins are integral elements in many constructed wetland stormwater treatment systems. Their design depends on the nature of the watershed and the loading calculations.

INFILTRATION

Infiltration, as opposed to either conveyance or detention, is designed to put water back into the ground, where it can be filtered and stored for future use or merely allowed to recharge the groundwater. In some regions, such as Maryland, Long Island, California, Florida, and parts of Arizona, this method is widely promoted by state regulators and might be called *recharge or retention.* It is superior to *detention,* which does not address problems of water quality, groundwater recharge, or conservation, and is being studied widely at present. Wherever local practice or law requires infiltration, the common underlying principle behind such ordinances is to reduce the flows and impacts downstream of the detention or retention facilities. An unstated principle that is being recognized more and more often is the desirability of creating infiltration opportunities for stormwater in order to recharge the groundwater. While this is especially important in the arid West, where underground water supplies are the only source of municipal water, it is also widely considered a sound management practice in other parts of the country. Although the principle of managing all stormwater on site and designing a site to allow maximum infiltration has been around for a long time, and practiced by landscape architects on a number of significant projects such as Woodlands, near Houston (1971), it has only recently become an accepted practice by engineers and regulatory agencies. Significantly, one of the main proponents of infiltration, water harvesting, and more aesthetic methods of managing stormwater onsite has been Bruce Ferguson, a professor of landscape architecture on the faculty of the University of Georgia (Ferguson, 1994).

WATER HARVESTING

The attempt to "harvest" stormwater and utilize it for landscape irrigation, one of the premises of the permaculture movement, is achievable only with difficulty under very limited conditions. There are severe limitations to the practicality of capturing, holding, and redirecting stormwater runoff that are related to the flow quantity, available space, site topography, and other factors. In areas subject to frequent high rainfall events, such as a 2-inch one-hour storm, the flows generated are significant enough to be highly erosive in many soil conditions, and therefore are capable of transporting and depositing large quantities of sediment in any basins or subsurface storage chambers. Therefore, interception or sedimentation devices of some type must be employed to minimize transport of sediment and other surface debris, such as bark mulch, which can quickly fill up any basin. As the highest rainfall often occurs during periods when the ground is already saturated, the best method of water harvesting that has a mechanism of controlled distribution is through subsurface storage in a tank or cistern structure. Continuing to direct runoff into landscaped swales or basins that are already flooded may lead to loss of any plant materials incapable of withstanding inundation. Unfortunately, unless a site possesses natural or recontoured topography with a natural slope from a storage chamber to the area requiring water, a submerged pump must be employed to pump the water to any distribution system, a use of energy at odds with other tenets of the sustainable, low-energy-use landscape. In addition, even if a particular landscape is in need of irrigation, the problem of designing a surface flow distribution system for stormwater runoff will always encounter difficulties related to even distribution without erosion and washouts.

That having been said, it is worthwhile to study the evolution of solar-powered pumps that to date have generally been capable of moving only fairly small flows, up to 5 gpm, without using very large collectors. In the very near future, concurrent with the development of increasingly efficient solar cells, the solar-powered pump may become more and more practical as an integral element in a sustainable landscape. Such systems for aeration and water recirculation are discussed in Chapter 7.

One of the most interesting demonstrations of water harvesting without any storage was designed by William Wenk of Wenk Associates, Inc., at his office building in Denver. The design captures all roof runoff water from the building and provides controlled distribution to water landscape plants carefully selected and placed in a sequence of moisture requirements ranging from high to low, each area allowing infiltration into the ground, with virtually no runoff allowed off the site except in high-intensity storm events (Figure 5.7a,b).

One example of water harvesting through the use of large storage cisterns is a large new art gallery in Santa Fe, New Mexico, built on a tight urban site within which all newly generated stormwater runoff from a 22,000-square-foot building and all paving must be detained. The design for this project entails capturing all roof runoff and holding it in two 50,000-gallon subsurface storage tanks. A submersible pump is utilized to provide water to an automatic underground irrigation system, redistributing the water to the landscape when needed. In addition, all parking area runoff is captured in a biofiltration swale that allows the water to filter

(*a*)

(*b*)

Figure 5.7 *Water harvesting, office of Wenk Associates, Denver (Craig Campbell).*

down into two buried 48-inch Hancor corrugated pipes, each 50 feet long. This runoff is then allowed to exfiltrate at a slow rate into the ground.

In general, subsurface storage of stormwater runoff is more practical as an infiltration method, through the use of large slotted or perforated pipes, than it is for water harvesting. Other methods that have been used to allow slow infiltration into the ground from subsurface storage rather than from unsightly, unvegetated detention basins are underground rock chambers wrapped with filter fabric to avoid clogging of the voids by fines. Stormwater is typically delivered to the chamber through a perforated pipe or from directly above through gravel, grass, or permeable pavement. The void volume of a rock-filled chamber must equal the sum of the volume of water to be handled and the volume of rock, which is equivalent to about 2.5 times the volume of water if the void space of the rock is 40 percent.

The *soft approach* to drainage problems involves minimum traditional engineering techniques along with the design of means of handling runoff that are sensitively integrated into an overall site landscape design. These methods may involve a combination of biofiltration swales, infiltration beds, and constructed wetlands that have proven their effectiveness in stormwater renovation. The combination of grassy swales, infiltration and sedimentation basins, and constructed wetlands is proving to be a sound solution in many situations for managing stormwater runoff.

VILLAGE HOMES

Bob Thayer and Tricia Westbrook (1990) prepared case studies of open drainage systems for residential communities for a conference of educators in landscape architecture that focused on examples in the Central Valley of California. The best example of an open drainage system is Village Homes in Davis, which happens to be Professor Thayer's chosen place of domicile. The author's conclusion is that given the relative costs of closed versus open drainage systems, the main opportunities for application of an ecologically sound open system design occur when a developer has a strong personal commitment to the ideals and values associated with this system. Thayer describes open or "natural" drainage as "the use of a system of *surface* topographic features such as swales, channels, and small ponds to collect and convey stormwater in a manner closely resembling natural watersheds." As he points out, conventional systems consist primarily of underground pipes, inlets, catch basins, and minimum overland flow of stormwater before it is captured; such systems have traditionally been designed by engineers, with the driving objectives being economics and protection of public health, safety, and welfare.

Village Homes is one of the most notable subdivisions in the country in terms of successful demonstration of conservation values. The designer, developer, and general contractor, Michael Corbett, located percolation ponds over areas with well-drained, gravelly soil. The same meticulous reading of the landscape that typified the earlier development of Woodlands, near Houston, Texas, by the firm of Wallace McHarg Roberts Todd was applied to the development of Village Homes. The drainage network at Village Homes has been described as a miniature watershed with dendritic systems of swales, a shallow retention pond, and check dams. Street drainage is channeled through notched curbs directly into the naturalistic swale system, which has been planted with a variety of vegetation. The entire system was designed to handle and absorb a ten-year storm event entirely on site, with any overflows being directed to the existing storm sewer in the street bordering Village Homes. In the eighteen years the community has existed, it has delivered overflow water into the conventional city system on only a few occasions.

OTHER EXAMPLES OF ON-SITE STORMWATER MANAGEMENT

Another example of a residential development utilizing surface flow stormwater management is Hybernia, Illinois, a development of 122 single-family houses on a 133-acre site in Highland Park north of Chicago. This site was designed around 16 acres of artificial ponds and a stream system that provide not only a unique setting but also stormwater handling through the conveyance of runoff in the artificial streams and detention of peak floods in the ponds.

Landscape architects are appropriately leading the effort to develop more natural means of handling stormwater in ways that promote groundwater recharge, as well as in assisting in removal of pollutants. There are now, and will be in the future, strong objections to such systems by many engineers and public officials who are still looking for cookbook designs and formulas for all installations dealing with either stormwater or wastewater. The development of closed systems to channel and convey stormwater runoff has been ubiquitous, but is totally understandable in areas where buildable land is in short supply. Essentially, conveying stormwater runoff into pipes requires the least amount of land, and any cost/benefit analyses undertaken to gauge the relative merits of piped versus open systems would not be capable of factoring in the ecological, water conservation, renovation, and aesthetic values associated with groundwater recharge, habitat, and visual qualities of open systems.

The long-range redevelopment plans for both the old Stapleton Airport, consisting of 1,400 acres, and the Lowry Air Force Base on 1,200 acres in Denver are based upon integration of surface flow stormwater systems in a series of interconnected swales, wetlands, and infiltration basins. A number of firms, such as Andropogon Associates and Wenk Associates, were involved in developing a sustainable landscape framework for redevelopment of these important urban spaces surrounded by areas already developed within the metropolitan Denver area (Fig. 5.8).

Bellevue, Washington—once considered a suburb of Seattle and now a prominent city in its own right–adopted a stormwater management system in the 1980s that incorporated a structure of natural local waterways, vegetated swales, and wetlands designed to protect the Kelsey Creek Fishery and to protect the community from flooding. In 1994, the city's Comprehensive Drainage Plan established additional objectives related to the CWA and set goals that included acquisition and rehabilitation of wetlands, preserving habitat for upland species, and working cooperatively with Parks and Community Services to integrate the natural open spaces that are components of the city's stormwater management system with park and recreational uses.

It is interesting to note that the majority of the constructed wetlands for wastewater treatment have been built in the western United States, while in the case of stormwater wetlands, the activity has been much more pronounced in the East, particularly in Florida and the mid-Atlantic area. The role of particular individuals, firms, and agencies that have taken the lead in specific areas of the country probably accounts for most of this distribution; however, I have a theory that may also explain the geography of current and past activity centers.

The entire western half of the United States is basically an arid zone, with the exception of the higher mountains and the Northwest on the west side of the

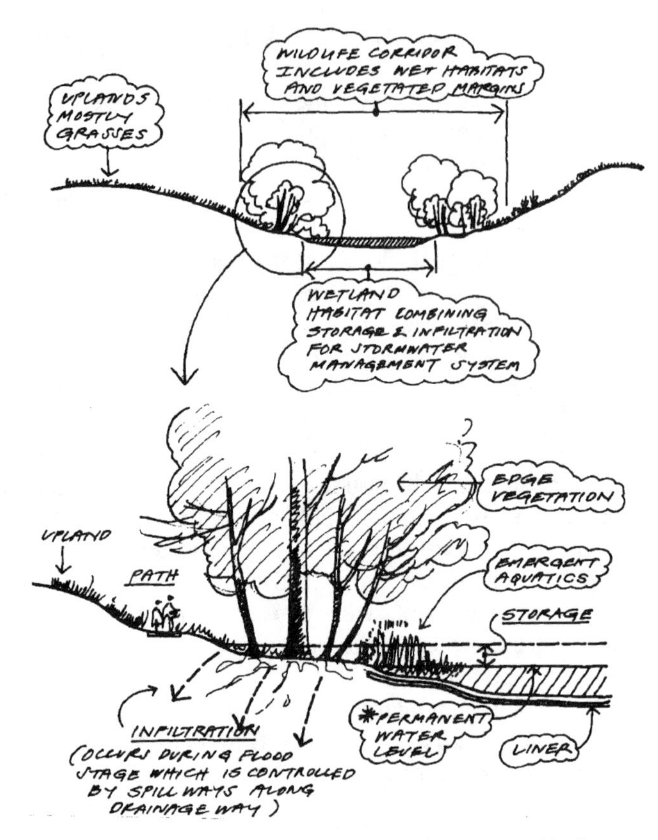

Figure 5.8 *Stapleton Airport storm drainage management concept (Andropogon Associates).*

Cascade Range. The appearance of riparian areas, marshes, and any lush landscape in general in most of the West provides what could be considered oases due to their visual contrast with otherwise brown settings. Therefore, from the beginning, constructed wetlands have had a very positive image in the West that coincides with the increased interest in permaculture, recycling of wastewater for irrigation, and other trends with considerable public support. Cattails, bulrush, reeds, yellow iris, and other aquatic plants literally glow within an arid setting and lack the negative image historically attached to "swamps" or wetlands generally. The rarity of these settings in the West compared to the East affects the public image of wetlands without question; and those individuals who have experienced the marvelous natural wetlands in parts of Nevada, eastern Oregon, New Mexico, and other parts of the arid West recognize the incredible beauty these settings provide.

In the case of stormwater wetlands, on the other hand, the problem of designing any basins capable of handling the typical high runoff events in much of the West while still supporting aquatic vegetation in the dry months is an extremely difficult one that has not yet, to my knowledge, been properly addressed. In much of the East and the Northwest, by contrast, soils with more organic matter permeability and more predictable moisture regimes spread over a longer period of the growing season allow for a number of options in the design of stormwater wetlands.

METROPOLITAN WASHINGTON COUNCIL OF GOVERNMENTS AND THOMAS SCHEULER: PROTOTYPE STORMWATER WETLAND DESIGNS

The Metropolitan Washington Council of Governments, along with the Maryland Department of Environment, has taken the lead in developing guidelines for stormwater wetlands in their region. Thomas Schueler of the Center for Watershed Protection has been responsible for several of the best reports on the subject (Schueler, 1992). Among the many admirable results of these efforts are the inclusion of wildlife habitat, the development of more sophisticated planting communities based upon varying water depths, and more attention to the aesthetics of such projects.

Because this region has developed the most sophisticated guidelines for stormwater wetlands, it is worthwhile to examine some of the principles developed by Schueler and others. These principles, while focused on the mid-Atlantic region, may have equal applicability elsewhere.

Schueler (1992) has described four basic designs for stormwater wetlands as follows:

1. ***Shallow marsh system.*** Based upon a sizable contributing watershed area, often in excess of 25 acres, the shallow marsh system has a large surface area requiring considerable space. It also requires a reliable base flow or groundwater supply to support emergent wetland plants (Figure 5.9).
2. ***Pond/wetland system.*** The pond/wetland system requires less space than the shallow marsh and consists of two separate cells; the first is a wet pond, and the second is a shallow marsh. The functions of the wet pond are to reduce

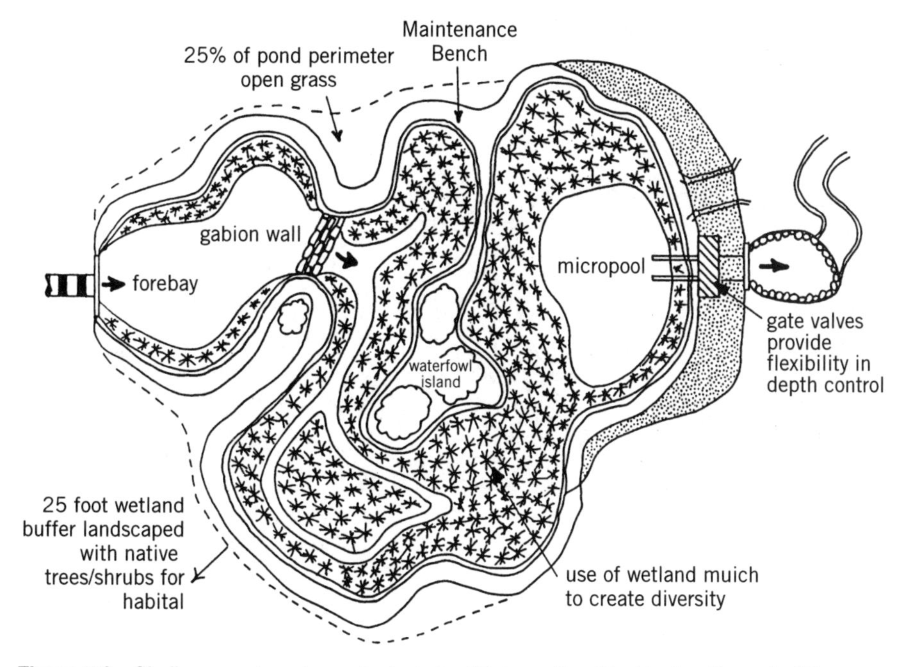

Figure 5.9 *Shallow marsh system—typical plan (Metropolitan Washington Council of Governments—Tom Schueler)*

incoming runoff velocity, trap sediments, and remove pollutants. As the majority of the treatment process is provided by the deeper pool rather than the shallow marsh, the system consumes less space than the shallow marsh system (Figure 5.10).

3. ***Extended detention (ED) wetland.*** In the ED system, additional temporary runoff storage is provided upstream of the shallow marsh, which enables this wetland system to consume less space. A vegetated zone along the side slopes of the ED wetland extends from the normal pool elevation to the maximum ED water surface elevation (Figure 5.11).
4. ***Pocket wetlands.*** These systems are appropriate for smaller sites (1 to 10 acres) that do not have a reliable source of baseflow and which thus experience widely fluctuating water levels. In some cases, pocket wetlands are excavated down to water table levels; in areas with lower groundwater levels, the wetlands are supported only by stormwater runoff. This type of wetland is probably the only one of the four that is applicable in the arid West, although a desirable modification to this design would be the addition of a sedimentation basin or forebay on the upstream side (Figure 5.12).

The selection of any of these designs is typically based upon three factors: contributing watershed area, available space, and the desired environmental function for the wetland. Schueler points out that stormwater wetlands are not equivalent to wetland mitigation or restoration projects in terms of species diversity and ecological function; they have the more limited goal of maximizing pollutant removal and creating "generic" wetland habitat. He also clarifies the difference between most natural wetlands, which depend upon groundwater levels, and stormwater wetlands, which are dominated by surface runoff in a "semitidal" hydroperiod

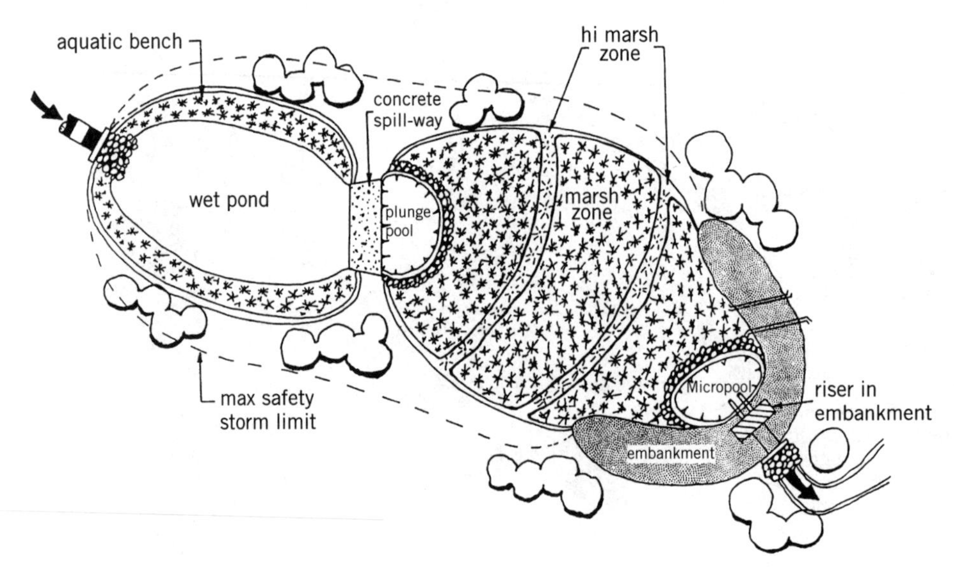

Figure 5.10 Pond/wetland system—typical plan (Metropolitan Washington Council of Governments—Tom Schueler).

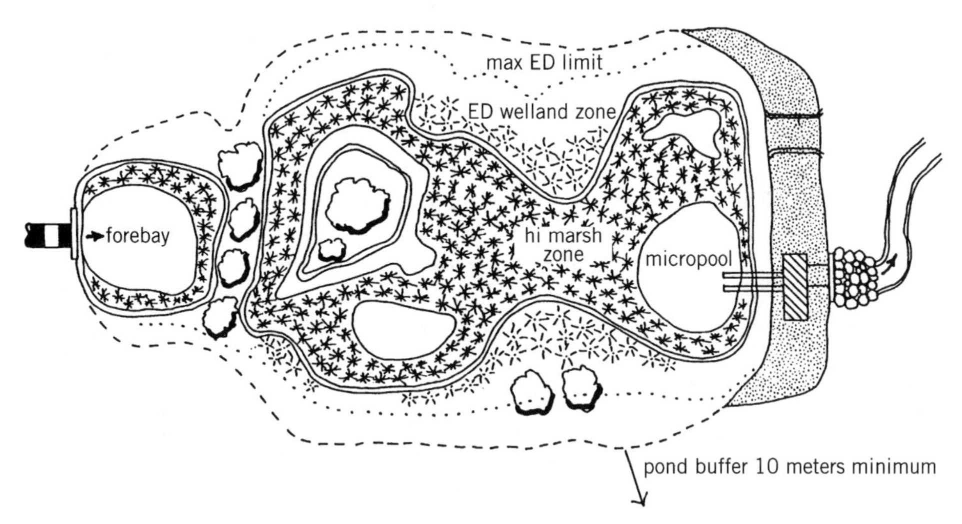

Figure 5.11 Extended detention system—typical plan (Metropolitan Washington council of Governments—Tom Schueler).

characterized by cyclic patterns of inundation and subsequent drawdown. Stormwater wetlands also typically experience greater sediment inputs than do natural wetlands. The designs and guidelines put forth by Schueler are far more sophisticated, in terms of the complexity of shapes, depths, variations in grade, and so on,

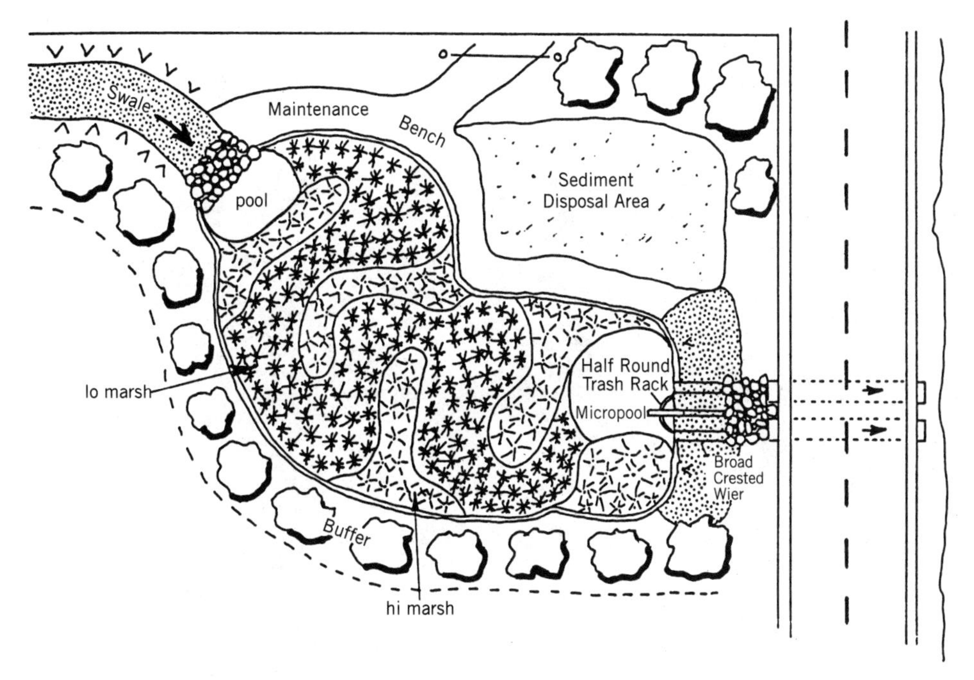

Figure 5.12 Pocket wetland—typical plan (Metropolitan Washington Council of Governments—Tom Schueler).

than past "engineered" wetlands with uniform depths, slopes, and plant communities. However, even with more sophisticated and ecologically sound designs, it is important to note that the morphology and cyclic hydroperiod of stormwater wetlands suffer from constraints of low plant life and wildlife, along with the need to address concerns of adjacent residents related to appearance, mosquito populations, safety, and other concerns. Stormwater wetlands are more likely to be dominated by five to ten species of wetland plants and are also more likely to provide an environment favorable to invasive or exotic species; natural wetlands exhibit more species diversity supported by existing prodigious seedbanks harboring dozens of species that may germinate only under particular conditions of moisture and temperature.

Dr. Kevin White at the University of Southern Alabama has utilized Schueler's stormwater treatment wetland models for the design of a demonstration project to treat approximately 2 acres of runoff from the University's golf course. The project was funded in cooperation with the Alabama Department of Environmental Management and the Gulf of Mexico Program. The project is illustrated under construction (Figure 5.13) and after planting (Figure 5.14).

SOME FUNCTIONAL DIFFERENCES IN STORMWATER AND WASTEWATER WETLANDS

In general, stormwater wetlands experience much greater sedimentation and adsorption of pollutants than wastewater wetlands with subsurface flow. Surface flow wetlands, on the other hand, perform similarly to stormwater wetlands, but with much less sediment input, if any. Algae are usually considered a problem in constructed wetlands for wastewater treatment, even though they provide oxygen, as well as taking up nutrients such as phosphorus and ammonia nitrogen. In the case

Figure 5.13 *Demonstration pocket stormwater treatment wetland in Alabama (Dr. Kevin White, University of Southern Alabama).*

Figure 5.14 *Demonstration pocket wetland after planting (Dr. Kevin White, University of Southern Alabama).*

of wastewater systems, the algae decompose once they die off and settle to the bottom; this process, in turn, will increase the BOD and possibly add to the suspended solids level of the effluent being monitored. Neither BOD nor suspended solids are of particular concern in a properly designed stormwater treatment system with adequate settling areas, and the outputs usually do not have to be monitored for these qualities. Therefore, algae can be of some benefit in any stormwater treatment wetlands, and indeed will probably become an unavoidable seasonal constituent of the system. Even the algal mats on the sediment surface of shallow wetlands are effective in removing nutrients.

POLLUTANT REMOVAL RATES FOR STORMWATER WETLANDS

A study of over twenty stormwater wetland systems that have been monitored indicates that most are capable of moderate to excellent pollutant removal. The following statements have been generally proven accurate:

- The removal rates for stormwater wetlands are similar to those for conventional ponds systems, such as dry EDs and wet ponds. In many cases, suspended solids removal rates are higher than those of conventional ponds, but phosphorus removal rates are more variable.

- ED ponds with wetlands typically perform better than ED pond without wetlands.
- The most reliable overall performance was achieved by pond-wetland systems (design 2) The permanent pond in this type of system pretreats runoff, reduces runoff velocities, and provides some pollutant removal before the runoff enters the shallow marsh. These systems also provided higher removal rates of nitrogen and phosphorus than other stormwater wetland designs.
- The performance of pocket wetlands has never been monitored. It is surmised that pocket wetlands will not be as reliable as the other stormwater wetland designs due to the interaction with groundwater and the lack of forebays.
- Several studies indicate that the performance of stormwater wetlands declines slightly during the nongrowing season and more strongly when the wetland plants die back in the fall, when they release a portion of the nutrients store in aboveground biomass. In addition, the lower temperatures reduce microbial and algal activity.
- Performance also declines if stormwater wetlands are covered with ice or receive snowmelt runoff.
- Some evidence exists that stormwater wetland performance increases as the wetland ages, at least in the first several years.
- In general, stormwater wetlands outperform natural wetlands of similar size, possibly due to the longer detention times designed into constructed systems.

Based upon the studies at the twenty stormwater wetland sites, Schueler projects the long-term pollutant removal rates for stormwater wetland in the mid-Atlantic region listed in Table 5.1.

EXAMPLES OF INTEGRATED STORMWATER WETLAND SYSTEMS

A number of landscape architectural firms, such as Tourbier & Walmsley of Philadelphia and New York City, Walker & Macy of Portland, Oregon, and Wenk

TABLE 5.1 Projected Long-term Pollutant Removal Rates for Stormwater Wetlands in the Mid-Atlantic Region

Pollutant	Removal Rate (%)
Total suspended solids	75%
Total phosphorous	45%
Total nitrogen	25%
Organic carbon	15%
Lead	75%
Zinc	50%
Bacteria	2 log reduction

Source: Schueler (1992).

Associates of Denver, along with several others, have been responsible for a number of interesting projects in which the objective of providing natural treatment of stormwater has been combined with park or open space usage and interpretive functions. Several of these projects will now be discussed.

Shop Creek

Wenk Associates of Denver has been in the forefront in experimenting with innovative stormwater management systems that both remove pollutants and create interesting and challenging landscape forms (Figures 5.15 and 5.16). Shop Creek lies within Cherry Creek State Park and contributed considerable phosphorous from a 550-acre watershed primarily covered by single-family homes into Cherry Creek Reservoir, a major recreation area. Working with Muller Engineering on hydrology and with Black and Veatch on water quality, Wenk developed a plan that incorporated a pond at the upper end of the park that collects and removes up to 50 percent of the phosphorus input, along with a series of constructed wetlands and sculptural drop structures. Shop Creek had deteriorated into a depressed channel with steep sides that were eroding every year. Wenk's design involved the installation of soil cement crescent-shaped terraces that provided small wetland basins to allow both treatment and infiltration. As flow rate slowed, the erosion rate was greatly reduced and the water quality improved.

Binford Lake/Butler Creek Greenway Master Plan

In 1992, Walker & Macy was commissioned to develop a master plan for 4 miles of greenway along Butler Creek and around Binford Lake for the city of Gresham, Oregon (Figure 5.17). The master plan developed for the area was based on a holistic approach that balanced water quality, wildlife, and recreational and edu-

Figure 5.15 *Shop Creek, Denver (William Wenk).*

Figure 5.16 *Shop Creek, Denver (William Wenk).*

cational concerns while providing a prototype for stormwater management and aesthetic enjoyment for the city. The plan incorporated biofiltration swales planted with specific grasses to intercept and filter stormwater runoff before it enters the lake; a boardwalk with interpretive signage; an enhanced wetland to filter pollutants and provide forage and shelter for wildlife; and a water quality pond to intercept stormwater fully designed with appropriate plantings for a fluctuating water level ranging from 6 inches to 2 feet. The interpretive signage not only described the project's goals and problem areas addressed by the plan, but also identified cautions regarding fertilizers and toxins and conveyed the concept of integrated pest management, along with the biological means of addressing water pollution.

High Desert Community, Albuquerque, New Mexico

Albuquerque, New Mexico, is one of the fastest-growing cities in the Southwest. As in Phoenix, Tucson, and Colorado Springs, most of its growth occurred during the 1950s and 1960s. Although the city receives only 7 to 8 inches of rainfall each year, it abuts a major mountain range—the Sandias—which soars more than a mile in height above the city. Significant rainfall events occur each year, many originating in the mountainous area, which receives considerably more precipitation. In the past, the city developed a number of large, concrete trapezoidal channels to handle major storm runoff, creating both visual blight and a dangerous conveyance that carries away people every year, some rescued and others not. Although new development is expected to detain water on site, the overall philosophy has been not to treat and infiltrate the stormwater, but merely to store it temporarily and allow it to enter storm drains, and ultimately the Rio Grande, at a slower rate. A refreshing alternative to this management philosophy, one that should become a model for future developments in the area, is represented by High Desert, a new community being developed on 1,200 acres lying in the foothills of the Sandia

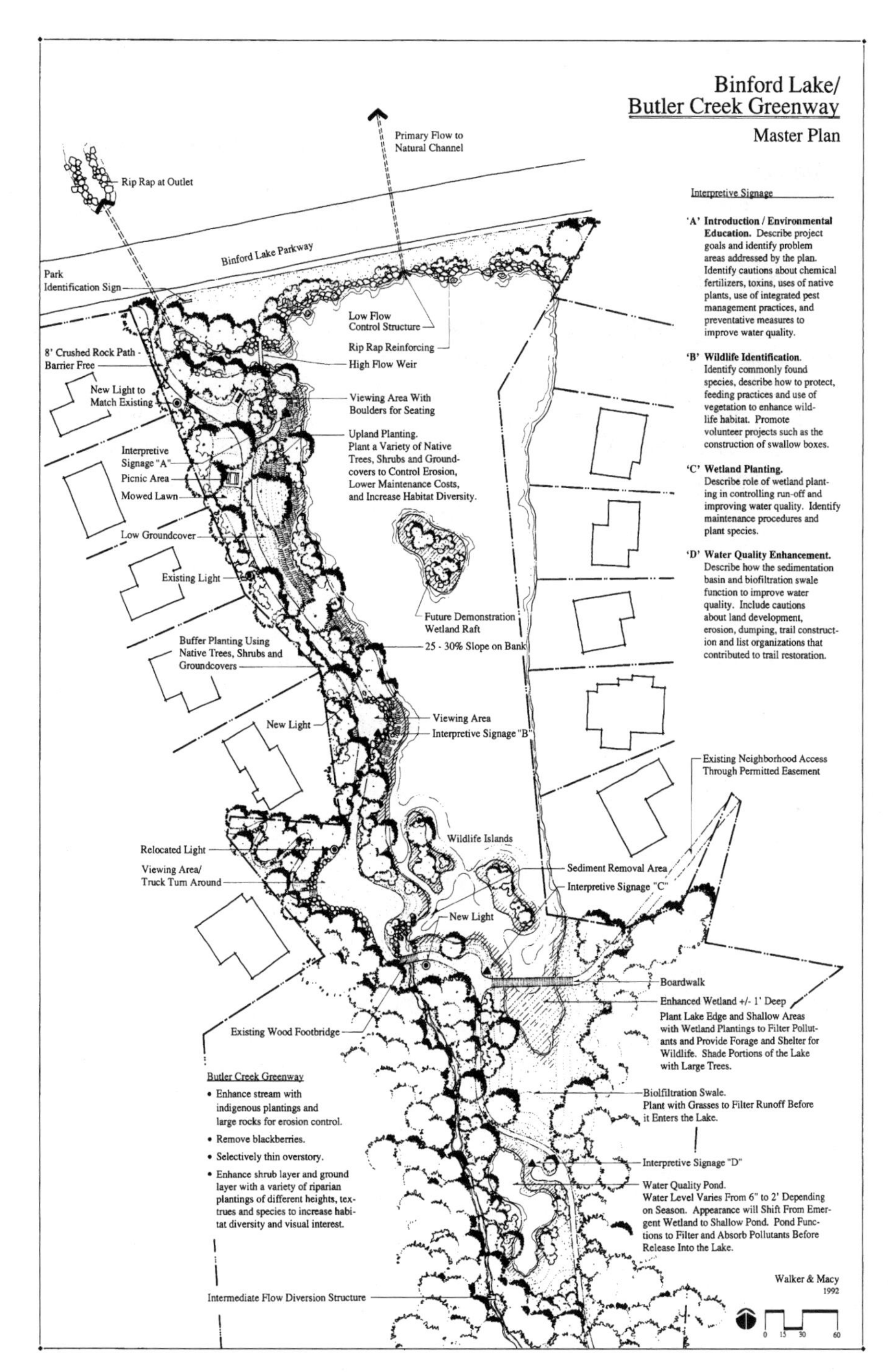

Figure 5.17 *Binford Lake (Walker Macy).*

range on the east side of Albuquerque. Rather than filling in the existing arroyos and installing buried stormwater drainage pipes, the development is planning an extensive water harvesting system that will utilize underground cisterns for storage and irrigation. All parking area and roof runoff will be filtered to eliminate sediments and contaminants. Guidelines for homeowners are provided to allow them to establish water harvesting system on their own lots.

Greenwood Urban Wetland, Orlando, Florida

The City of Orlando adopted innovative and stringent ordinances in the 1980s that regulate the management of surface water runoff quality as well as quantity, preceding federal regulations on management of stormwater pollution. In addition, Florida's Department of Environmental Regulation developed criteria for the design of new "wet" detention systems that required, for the Orlando area, 1 inch of runoff storage above the level of any permanent pool and a maximum discharge of one-half of the runoff volume in the first 60 hours following an event. They also recommend a littoral shelf sloped at 6:1 or flatter out to a water depth of 2 to 2.5 feet and a maximum of 70 percent open water. The remainder, the littoral shelf, is to be established with aquatic vegetation. Facilities that are sources of oil and grease contamination are required to include skimmers or other means of preventing these substances from leaving the facility. Pretreatment in the form of landscaped retention areas or biofiltration swales is recommended for incorporation into the stormwater treatment system to provide infiltration and treatment of the first half inch of runoff.

Due to the high visibility of the site, Orlando decided to develop a project that not only protected the quality of its groundwater supply, but also provided an aesthetically pleasing stormwater facility. The concept of creating a parklike environment with walkways, footbridges, and native landscaping was therefore one of the central objectives of the stormwater treatment project. The engineering firm of Dyer, Riddle, Mills, & Precourt of Orlando was engaged by the City of Orlando to design the facility.

Information signs were developed to describe the various plant communities, the stormwater treatment process, and the water conservation concepts (Figure 5.18). Three different plant communities were established as part of the project. Two of the systems were designed for water quality treatment and the third to create the native, upland forest for the park. Over 850 trees and 82,000 wetland plants were installed

In a follow-up study of the effectiveness of this system for pollutant removal funded in part by the U.S. EPA (McCann and Olson, 1994), the system performed very well in lowering TSS and phosphorus, but the wetland was determined not to be effective in removing nitrogen; only ammonia and nitrite levels were slightly reduced, from 8 to 10 percent. However, as this system experiences considerable inflow from groundwater, which cannot be monitored, it was assumed that groundwater was a significant source of nitrogen into the system. Removal of metals was much better, with zinc removal efficiency at 68.9 percent; lead at 59.7 percent; copper at 58.9 percent; and cadmium, at an extremely low rate of loading, at undetectable levels.

Figure 5.18 *Greenwood Urban Wetland (FDER).*

Lake Matthews Watershed, Riverside County, California

John Tettemer & Associates of Costa Mesa, California, designed a water quality–wetlands demonstration project in the Lake Matthews watershed in Riverside, California, that is intended to provide data on water quality and the use of wetlands for water quality purposes in the Southwest. The intention of the project is to improve the water quality of storm runoff before it reaches Lake Matthews, which supplies drinking water for more than 15 million Southern Californians.

The project incorporates a wide range of BMPs, both structural and nonstructural, along with three water quality wetland facilities located at strategic hydrologic drainage collection points to provide treatment for first-flush stormwater runoff. Also included in the project is a first-flush diversion structure and water quality pond, a sediment basin, a detention basin, and five sediment wetland basins. An existing wetland area is being enhanced to provide a wetland corridor designed to convey a 100-year flood in addition to providing wildlife habitat expansion. This feature, known as the Cajalco Creek wetlands corridor, consists of fragmented wetlands that are being consolidated to provide a continuous riparian corridor throughout the central area of the adjacent watershed. Other wetlands, including several wetland basins planted with emergent plants, will be located off-stream, where they will be supplied with water with a first-flush diversion structure. Other stormwater detention ponds have been designed to hold irrigation water for a golf course irrigation system.

David R. Bingam (Kent, 1994) pointed out that although there is no general agreement on specific design criteria, there appears to be a consensus on what should be considered or evaluated during the design. Bingham compared the guidelines currently used in Florida, Maryland, and Washington for constructed wetlands for stormwater treatment. The Washington State Department of Ecology recommends an oil control device or separator prior to the forebay; in Florida, it is recommended before the outlet. The use of constructed wetlands for stormwater treatment is still an emerging technology, and much must be learned about the design and performance of different systems under a variety of climatic and site conditions. The potential for establishing an entire new community based upon integration of surface management of stormwater, with a network of constructed wetlands and ponds, is enormous. Such a project may be carried out on a major scale with the redevelopment of both the Lowry Air Base and the old Stapleton Airport lands in Denver for housing, park, and retail development over the next decade.

6

Single-Family Residential Systems

We are the children of our landscape; it dictates behavior and even thought in the measure to which we are responsive to it.

—Lawrence George Durrell (*Justine,* 1957, pt. 1)

Peter Warshall (1979) wrote a wonderful little book that commented on the attitudes regarding waste disposal that evolved in western cultures concurrent with industrialization. He says: "Instead of eating, defecating onto the ground, fertilizing plants with feces, and eating again, we simply reach behind our backs and pull a little chromium lever. Instead of defecating into the earth, we sit on a toilet filled with good drinking water which comes from some unknown river and, after flushing, goes to some unknown destination" (p. 1). He correctly points out that wastes are misplaced natural resources that cannot be disposed of but must ultimately reenter the nutrient and water cycles of our planet. Our current system of centralized treatment systems, rather than recycling nutrients into the soil in a usable fashion, merely attempts to remove and relocate them to places where they are disposed of in streams and lakes, where they are not needed and create environmental problems. The move toward big sewers, as Warshall points out, was supported by health officials, engineering firms, water companies, and real estate developers, all of which realized benefits from centralized treatment systems. The cost of centralized collection systems to sewer newly developed areas or older unsewered areas is extremely high and creates a situation that is in direct opposition to the principles of sustainability. Typically, when cost comparisons have been made, centralized treatment systems have been found to cost 2.4 time as much as decentralized systems that cluster up to five houses on one drainfield and 3.0 times as much as individual home-site systems.

THE CASE FOR ON-SITE TREATMENT

Warshall argues that on-site sewage treatment pollutes less, costs less, uses less of our energy resources, and is less of a health hazard than centralized sewage treatment. There are obviously some site conditions related to high groundwater, rock or impermeable soils, topography, and other factors that are unsuitable for on-site treatment. Although millions of dollars have been provided to centralized treatment plants, there has been no sizable source of financial support to upgrade and study home-site treatment systems. Most problems identified with home treatment systems simply result from the lack of public information on proper installation and maintenance techniques, a situation some states are attempting to rectify by establishing agencies to provide assistance and by monitoring such system. There are an estimated 30 million on-site wastewater treatment systems in the United States, and approximately 57 percent of the housing units in Vermont use such systems. An estimated 25 percent of the U.S. population has been using on-site systems since 1970. This percentage may increase with the movement of people away from cities to rural areas.

Dr. Anish Jantrania, Technical Services Engineer with the Virginia Department of Health, states the case for these systems (Jantrania, 1998):

> Onsite systems are a viable way to treat and dispose of wastewater. They are no longer considered temporary solutions until a community can be sewered. Centralized wastewater management is very expensive, both economically and environmentally. From an economic standpoint, common sense tells you that it is very expensive to collect wastewater from a thinly populated area and take it to another place to treat and dispose of it. From an environmental perspective, it makes sense that if you are getting your water from the ground, you should treat it and return it to the ground rather than discharging it off the site to surface waters. It is important to keep our groundwater and surface water supplies in balance.

points out that a whole series of new technologies allow systems to be developed for any site. In addition, there is increasing recognition of the environmental benefits derived from on-site wastewater treatment, along with the reuse of graywater. California officially adopted statewide residential graywater reuse standards in 1994 in response to its 1992 law legalizing residential graywater for subsurface landscape irrigation. Several cities are conducting research on water savings resulting from graywater reuse, which may achieve a "preferred" management status creating a movement toward requiring graywater plumbing in new construction. Recycling of graywater greatly reduces the overall wastewater flow from a residence and increases the overall load by eliminating the dilution provided by sink, bath/shower, and laundry water. In general, it is not advisable to connect a kitchen sink or dishwasher to a graywater system due to the amount of fat, oil, and grease they contain, which is difficult to filter and may clog distribution pipes.

There are special considerations to keep in mind when designing a residential constructed wetland system in combination with a graywater recycling system. Until very recently, it was illegal even in the State of California, which has suffered severely from droughts, to install graywater systems; but attitudes are changing nationwide, and entire subdivisions are now being built with separate graywater plumbing systems. In 1995, the International Plumbing Code was revised to allow for the utilization of graywater recycling systems.

When we consider constructed wetlands and their potential for landscape integration, the single-family residence offers unique possibilities for weaving such features into the surrounding landscape in a functional and imaginative way. Constructed wetlands for individual residences are common in some states and rare in others. For example, according to local regulatory agencies, at least 1,000 of them have been installed in Kentucky; 100 in Arkansas; 40 in Louisiana; 200 in New Mexico; and smaller numbers in Alabama, Colorado, Missouri, Texas, Virginia, and West Virginia. In many, if not most, other states, constructed wetlands for treating the wastewater of single-family residences are nonexistent. Various terms, in addition to *constructed wetlands,* have been utilized to describe these systems; in the Southeast the terms *rock filter* or *rock plant filter* are commonly employed and are used interchangeably in this chapter with *constructed wetlands*. Little effort has been at uniform monitoring of residential systems outside of Kentucky and Arkansas.

In some specific areas, there are several site conditions that dictate reasons to consider the benefits of constructed wetlands for a single-family residence. The typical situation to begin with is a location outside the boundaries of a centralized sewer collection system, along with the requisite requirement to install a septic tank and leach field on the property. If the property has either a high groundwater table or shallow rock, the existing site conditions are unsuitable for a typical leach field. In some states, constructed wetlands are allowed for such residential sites only if they are at least 3 acres in size. These states are operating on the assumption that there will be a surface discharge of the treated effluent that may enter another property.

A fairly common rationale for considering a constructed wetland derives from an owner's or developer's commitment to demonstrate concern over groundwater contamination and the desire to utilize natural treatment methods as part of an overall philosophy of life. This philosophical support often produces the most imaginative systems, usually well integrated into an overall landscape (Figure 6.1).

SEPTIC TANKS—THE FIRST STEP

What must be understood here is that even with a constructed wetland installation, primary treatment is still required, as provided by a septic tank that still requires normal periodic maintenance. In addition, it must be stressed that with a constructed wetland system, there will be an effluent stream that will still require disposal. The options, however, include infiltration beds that operate similarly to a leach field, but with higher-quality effluent; an evaporation pond; a decorative pond; or an irrigation vault or pond used to water ornamental plantings, fruit trees, and so on.

All of these options have been utilized at one location or another in the Southwest. The other region that has quite a few residential-scale constructed wetland systems in place is the Southeast, particularly Alabama and Tennessee, thanks to the influences of Billy Wolverton and to the efforts of the TVA.

Septic tanks provide the first step in the treatment of residential or other small-scale wastewater. They are biological digesters that provide an ideal environment

Figure 6.1 *Lusk residence, New Mexico (Craig Campbell).*

for both anaerobic bacteria and some *facultative* microorganisms, which are capable of living either in the presence or absence of oxygen. Each type of bacteria plays a role in feeding on the solids that settle out of the liquid within the tank. Septic tanks, when properly designed and maintained, provide a habitat for anaerobic bacteria that are essential in the initial processes involved in treating wastewater. Much misunderstanding exists among politicians, regulators, and the general public about the nature and effectiveness of septic tanks. In various parts of the country, this confusion results in periodic bans on septic tanks in new developments in favor of package treatment plants or other centralized systems. Most failed septic tank systems are ones that were poorly designed or constructed, improperly maintained, too old, or overloaded. Not enough attention has been paid to quality control in the septic tank fabrication industry, and leaking tanks are common. In addition, many jurisdictions fail to apply rigorous standards to the ability of soils to percolate properly, and they allow leach beds in areas with extremely poor percolation. As previously mentioned, many septic tank improvements, retrofits, and additional treatment units ranging from intermittent sand filters, to trickling filters within the tanks, to constructed wetland components can be integrated into a septic tank-based on-site treatment system to provide levels of treatment rarely achievable at a centralized treatment facility. In recognition of the benefits of well-managed on-site systems, along with the increased knowledge and improved technologies now available, some communities and counties have adopted ordinances requiring all septic systems within their jurisdiction to be brought up to new standards, with no "grandfather" clauses allowing any systems to avoid being upgraded.

There has been much misinformation regarding septic tank processes and additives that claimed to enhance septic tank performance, clear drain fields, and so on. Many different products, usually based upon a mixture of specific bacteria and enzymes, have been marketed to enhance the performance of septic tanks and drain fields. The active research going on in the field of bioremediation has provided the basis for newer products that may be more effective than those previously on the market. One of the few reviews of scientific data on the subject was undertaken by Prof. Kate M. Scow, of the University of California at Davis. In her report published in 1994, entitled *The Efficacy and Environmental Impact of Biological Additives to Septic Systems,* she reported her conclusions that there is no evidence that biological additives have an adverse impact on septic tanks or the environment; that such additives can enhance the performance of septic systems; and that older studies on additives may not be relevant because the formulations of products have changed. It is interesting to note that the problem that formed the basis for Professor Scow's research was the passage of a law by the State of Washington that banned the use, sale, and distribution of unapproved additives to on-site sewage disposal systems. Although the law was clearly aimed at chemical additives such as solvents and caustic substances, biological additives were also covered.

In a related study, *Assessment of the Effects of Household Chemicals Upon Individual Septic Tank Performance,* in 1987, M. A. Gross of the Graduate Institute of Technology of the University of Arkansas at Little Rock tested the amount of household chemicals required to destroy the bacterial population in an individual domestic septic tank. The tests indicated that an excessive amount of any of the cleansers or disinfectants applied in a slug loading (all at once) destroyed the bacteria, but that after normal septic system usage, the bacterial population returned to its original concentration within hours. Drain openers appear to be the most toxic to bacteria. In general, any cleaning agents, disinfectants, or drain openers that adversely affect the bacterial population in a septic tank may also affect plants in a constructed wetland. For that reason, and to support the general philosophy of using less toxic chemicals in homes, schools, and the workplace, the utilization of constructed wetlands should be accompanied by a manual or guide designed to persuade those responsible for cleaning operations to use alternative cleansers such as white vinegar. A good substitute for chemical drain openers is a mixture of one-fourth cup of baking soda, one-half cup of vinegar, and 1 gallon of boiling water.

The most probable reasons for the negative image of septic tank and leach field systems are the lack of design standards for particular site conditions and the lack of understanding regarding maintenance. Warshall (1979) discusses the development of local agencies such as on-site waste management districts, which keep records of all septic tank/drainfield locations, dates when they were pumped, types of soils in the district, and so on, and develop guidelines for how new on-site systems should be designed and built. The districts typically have the right to inspect septic tanks periodically, require pumping if necessary, and require repairs to systems that might have the potential to cause health or pollution problems. In this manner, if private home-site systems were given the attention they deserve, there would be far fewer failing systems, which have given such installations a bad reputation in some areas. Such districts have been established in California and Wisconsin, some of which have undertaken responsibility for managing all on-site systems in their jurisdiction. There are both community and countywide management districts in several counties in California where monthly fees are charged or fees are levied onto the property taxes.

CONSTRUCTED WETLANDS FOR THE RESIDENTIAL PROPERTY

There are several different alternatives for residential on-site treatment of wastewater, ranging from composting toilets to evaporation mounds. There are also alternatives for both new and existing residential developments, assuming that the land is available, to collect wastewater from septic tanks in clusters of two to five or more houses and develop a constructed wetland to achieve further treatment before disposal, either subsurface or in an evaporation pond, which can be developed as an attractive wildlife habitat feature.

The original impetus toward application of constructed wetland systems at a single-family residence level originated with the TVA in Knoxville, Tennessee, which, as previously mentioned, had developed an active unit to study and promote constructed wetlands for community and other larger systems, as well as for acid mine waste renovations. In 1991, the TVA produced the first manual directed at individual home owners, consultants, and contractors interested in building constructed wetlands for small systems down to the size of the single family. Following up on their experience with various problems, the TVA issued an updated guide in 1993 that represents an admirable effort at technology transfer. Prepared by Gerald R. Steiner and James T. Watson, the report is entitled *General Design, Construction, and Operation Guidelines: Constructed Wetlands Wastewater Treatment Systems for Small Users Including Individual Residences.* "Small" systems are considered by TVA to be systems treating flows of 20,000 gallons per day or less. The TVA reports that the smallest system they have designed has been for 83 gallons per day for a one-bedroom house with limited wastewater. The report covers a description of pretreatment, loading, design, vegetation, and operation and maintenance, and also includes several construction details. The TVA actively encouraged feedback on both positive and negative experiences to allow them to update their guidelines (Figure 6.2).

Figures 6.3 and 6.4 illustrate a residential constructed wetland system in Alabama modeled on the TVA guidelines that was designed and installed by the Department of Civil Engineering at the University of Southern Alabama. Dr. Kevin White of the department has developed a constructed wetland Web site and has

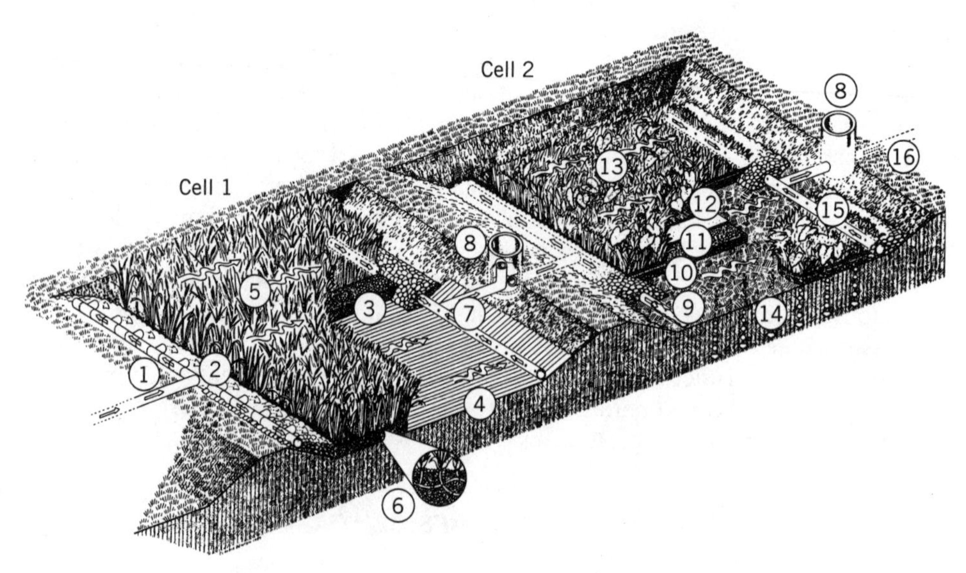

Figure 6.2 *Constructed wetland (courtesy of the TVA).*

Figure 6.3 *Residential constructed wetland, Alabama (Dr. Kevin White, University of Southern Alabama).*

worked with the State of Alabama Health Department and the U.S. EPA Gulf of Mexico Program on several pilot projects involving both wastewater and stormwater treatment wetlands.

The TVA's guidelines for residential constructed wetland systems suggests a one-cell system for areas where wastewater will not percolate into the ground and a two-cell system in series where soil will marginally percolate treated wastewater.

Figure 6.4 *Residential constructed wetland, Alabama (Dr. Kevin White, University of Southern Alabama).*

The first cell is lined and the second cell is unlined to allow percolation and minimize surface discharges. The calculation of wetland size for cold areas is suggested as 1.3 square feet per gallon per day. In other words, for a typical two-bedroom house, the required constructed wetland would need to be between 300 and 400 square feet, depending upon the local rates of calculated flow. Berms to contain the liner and earth cover must be allowed for in addition to the size of the constructed wetland; this might add an additional 200 to 300 square feet to the required space, depending upon the details utilized. When rocks are used to define a constructed wetland edge, less space is required and there is a more pleasing visual effect than that achieved with an unplanted berm. In the Southwest, an earth berm is likely to remain bare, while in more humid parts of the country, it would be practical to cover the berm with appropriate groundcover plantings or grasses.

The TVA has developed a detail for a constructed wetland cell wall consisting of plywood, railroad ties, and 2 × 4s that may be appropriate in the Southeast and relatively easy to install but that may not be suitable in drier parts of the country (Figure 6.5). There is evidence that the TVA's recommendations for residential system sizing were too small, resulting in poor performance, along with some criticism of the standards as unsuitable in areas with colder climates; in areas where nitrogen removal is important; and for systems larger than single-family residences in scale. Unfortunately, similar sizing recommendations were later repeated by Louisiana and Arkansas, resulting in similar problems in those states, which precipitated a temporary moratorium on such systems. The report calls attention to the benefits of installing a filter on the effluent side of the septic tank to further reduce solids and organic load to the constructed wetland system; such installations are cost effective and low in maintenance.

The term *hydraulic loading* refers to the quantity of effluent in gallons per day delivered to the constructed wetland. This is calculated on the basis of number of

Figure 6.5 *Residential system nearing completion, Alabama (Dr. Kevin White, University of Southern Alabama).*

bedrooms or fixtures at rates that are established independently by each state. Typical rates are 120 to 150 gallons per day per bedroom, but local experience in terms of actual usage may vary considerably from the average national figure. Many long-term residents in arid zones, especially where water rates are high, tend to be somewhat more conservative in their use of water. Almost all new constructions in such areas utilize low-flow toilets and some include on-demand water heaters, both of which tend to lower the average water use per person. *Organic loading* refers to the BOD calculated in pounds per day. Average daily organic loading per person would be approximately 0.17 pounds BOD_5 per person per day to the septic tank, which would be expected to reduce the load by 50 percent before delivering effluent to the constructed wetland. Therefore, the TVA recommends a rate of 0.085 BOD_5 as a reasonable calculation for the loading to a wetland system. (The "5" in the BOD measurement refers to a five-day period for measurement purposes) They also point out that if two septic tanks in a series are used, or a effluent filter is installed, the total organic load reductions may be 70 percent or greater. There have been numerous experimental modifications and additions to septic tanks to improve the treatment levels, particularly to lower the nitrate form of nitrogen. Some of these involve aerated cells; others involve small trickling filters retrofitted to receive recirculated effluent from the discharge end and provide efficient nitrification of ammonia to nitrate. Upflow filters are then used to denitrifry, or remove nitrate from, the effluent. Constructed wetlands have been shown to provide excellent denitrification, and further experiments will undoubtedly be undertaken to test the combination of septic tanks, small trickling filters, and constructed wetlands or rock plant filters to produce high-quality residential wastewater treatment.

Other guides or manuals for on-site plant rock filters intended for individual homes have been prepared and distributed by the Arkansas Department of Health, the Louisiana Department of Health, and the Lexington–Fayette County Health Department in Kentucky.

The State of Louisiana, requires two or three septic tanks in series or a single tank divided into no fewer than two compartments, except for a one-bedroom residence. The design guidelines issued by Louisiana are given in Table 6.1.

The plant filter must be designed to maintain a liquid depth of 12 inches, even during periods of no flow, and a total rock depth of 18 inches is recommended. The State of Arkansas issued similar guidelines, revised in 1991, as "Interim Design Criteria for Rock Plant Filters," with recommended filter sizes approximating those of Louisiana. However, the filter sizes are somewhat smaller for the higher

TABLE 6.1 Louisiana's Guidelines for Residential Constructed Wetlands

No. of Bedrooms	Septic Tank Size (Gal)	Rock-Plant Filter (Ft^3)	3-ft and 6-ft Width/Length
1 (250 gpd)	600 (single compartment)	100	3′ × 33.3 or 6′ × 16.6′
2 (300 gpd)	750 (2–3 compartments)	200	3′ × 66.6′ or 6′ × 33.3′
3 (400 gpd)	1,000	300	3′ × 100′ or 6′ × 50′
4 (500 gpd)	1,250	400	3′ × 133.3 or 6′ × 66.6′
5 (600 gpd)	1,500	450	3′ × 150′ or 6′ × 75′
Over 5 (700 gpd)	2,000	500	3′ × 166.6′ or 6′ × 83.3′

gpd = gallons per day.

flows from a four-bedroom house than Louisiana's, even though the calculations are based upon a flow of 600 gallons per day for a 4-bedroom house compared with 500 gallons per day in Louisiana. In general, most of Louisiana has milder wintertime temperatures than most of Arkansas; and one would expect that, as data are collected, the minimum filter sizes required in Arkansas would increase or those applied in Louisiana would decrease to reflect these differences.

As previously mentioned, there is considerable evidence that the sizing recommendations published in the past by the TVA and the states of Louisiana and Arkansas are inadequate and need revision. At present, when soil or other conditions limit the use of subsurface disposal system, Arkansas requires a minimum 3-acre property prior to approving the use of rock plant filters to avoid any possibility of a surface discharge off site that might violate NPDES discharge standards. It is entirely acceptable, however, for homeowners in a site approved for a septic tank and leach field to install a rock plant filter at their own discretion.

In *An Analysis of Four Rock Plant Filters in Fayette County, Kentucky 1992–93*, covering a twelve-month period, the Lexington–Fayette County Health Department concluded the following:

1. The plant rock filter is a very effective means of treating septic tank effluent composed of individual household liquid waste.
2. The plant rock filter provides good treatment in its initial stages, and the degree of treatment continues to improve for three years or more as the aquatic plants continue to mature.
3. The tests indicate that adequate treatment is provided during all months of the year and in both summer and winter. The rock filters tested appear to be too immature to allow a true comparison of summer and winter treatment.
4. To determine the full effects of the plant rock filter on the treatment of wastewater, tests should be conducted on rock filters with fully mature plants. This should be done when the plants are four to six years old.
5. The plant rock filter tested with plants two to three years old was compared to the Kentucky Sewage Treatment Discharge Permit Standards for the three package sewage treatment plants and the Fayette County Board of Health Standards for Sewage Treatment Plants. Except for dissolved oxygen, the plant rock filter met these standards.
6. The plant rock filter tested with plants between two and three years old has effluent with fecal coliform counts that are considerably below surface water runoff from three urban sites in Fayette County.
7. The location of the outlet at the bottom of the rock filter provides water quality similar to that provided by the outlet at the top along the flow line of the rock filter except for dissolved oxygen. Dissolved oxygen is significantly greater at the top of the rock filter than at the bottom. This situation could change either way as the rock filter matures. The outlet at the top continues to have the physical advantages of better elevation and of avoiding leaks in the plastic liner when a hole is cut at the bottom.
8. Because the construction of the rock filter allow no wastewater contact with soil until after it has been treated, this is one of the most pollution-free methods of treating wastewater.

Why should a homeowner or developer consider installing a constructed wetland system? If the site conditions of high groundwater or shallow rock, previously mentioned, are present, then there are few alternatives that are acceptable to local regulators. If the homeowner is interested in the possibility of reusing wastewater to water ornamental plants or for a pond, then the attractions are obvious.

In the Southeast, a number of residential wetlands have been constructed utilizing canna lilies in the lined gravel trenches, thus creating a wonderful field of flowers during much of the summer in place of an invisible leach field that contributes nothing to the beauty of the grounds. While most rock filters installed at individual residences have been constructed in a simple rectangular shape, one of the installations tested in Fayette County is circular, is located in a driveway circle, and is a landscape focal point. The possibilities for integrating rock filters in residential landscapes are fascinating and have only begun to be explored.

It is fair to point out that the most successful and interesting applications of constructed wetland systems at the single-family residence level are seen in situations where the homeowner is a dedicated gardener, tinkerer, and do-it-yourself type. This is not to imply that these systems should never be considered except otherwise, only that, to gain the maximum benefit from a constructed wetland system, one must be willing to experiment, sometimes face regulatory challenges, and have a gardener's sense of what must be done in terms of plant maintenance, replacement, and other characteristics of homeowners with interesting gardens in general. After all, a constructed wetland is really an extension of the landscape that needs to be well integrated with other elements of any residential landscape.

Ponds, as special features in the landscape, are addressed in Chapter 7.

It must be stated that the data obtained from residential constructed wetlands are extremely limited and not always reliable. Due to the lack of performance data, some states have imposed temporary moratoriums on further installations of constructed wetland until more data are collected. Performance under one set of climatic conditions cannot be used to predict performance under another climatic regime; and it is quite likely that the winter temperatures and the length of the growing season in the southern states and the temperate West Coast will allow for better performance than achieved by systems of the same size installed in colder climates.

PLANTS

There is such variation in climate, soils, pH, and other conditions from one part of the country to another that it is difficult to develop lists of aquatic plants with universal application. Other than common cattail, reed, and bulrush species, most other aquatic plants have geographic limitations that are best researched with local nurseries and aquatic specialists. In addition, many excellent books have been published over the last decade or so on water gardening in response to the increased interest in these residential features. Some of them are discussed in Chapter 7.

Residential constructed wetlands offer the possibility of combining plants whose main function is to establish vigorous roots, and thus provide the main treatment function, and plants along the edges or borders of the wetland whose primary value is ornamental. Since residential systems usually involve shallower beds than larger subsurface flow constructed wetland systems, there are greater opportunities to incorporate a wider variety of aquatic plants. Highly invasive plants may tend to dominate a residential system and should be avoided unless a single aggressive species is desired for a uniform effect. Some states provide a list of approved plants for plant rock filters; any plants not on the list must be approved before they can be used.

One of the best ornamental aquatic plants, which seems to be extremely hardy and adaptable to many climates, is the yellow iris (*Iris pseudocorus*), a fairly large, bulky plant that is excellent for borders or clumps in massing areas. It grows to about 2 to 3 feet, with roots penetrating about 10 inches, and produces yellow flowers in the spring. Like all irises, the yellow iris requires considerable sun in order to thrive. Blue iris (*iris versicolor*) is a slightly smaller plant with narrow, graceful foliage and roots penetrating to about 8 inches. Many other aquatic plants have also been used for residential constructed wetland ornamental borders, such as flowering rush (*Butomus umbellatus*). Golden club (*Orontium aquaticum*), horsetail (*Equisetum hyemale*), various fern species, graceful cattail (*Typha laxmannii*), narrow-leafed cattail (*Typha angustifolia*), reed mace cattail (*Typha stenophylla*), bog arum (*Calla palustris*), butter cup (*Ranunculus spp.*), arum lily (*Zantedeschia aethiopica*), bog bean (*Menyanthes trifoliata*), arrow head (*Sagittaria latifolia* and other species), marsh marigold (*Caltha palustris*), pickerel rush (*Pontederia cordata*), sweet flag (*Acorus calamus*), and water primrose (*Ludwigia* spp.) are only a few. Another attractive plant native to the Southeast, which has been utilized in some rock filters, is the southern wild rice (*Zizanias miliacea*), a perennial with rhizomatous roots that attracts birds. Many species and varieties of grasses and sedges are also appropriate for margins of wetlands.

Aquatic plants need to be kept in water until planting, and preferably planted the day they are obtained. It is best to separate plant clumps into separate plants, trim back the roots to stimulate new shoots, and wash the roots completely free of soil.

A discussion of two residential on-site constructed wetland treatment systems follows.

LUSK RESIDENCE, NEW MEXICO

One of the best examples in the country of a well-conceived, beautifully integrated constructed wetland wastewater and pond treatment system at the single-family residence level was created by Paul Lusk, a professor in the Department of Architecture at the University of New Mexico, at his property in the South Valley, a historic semirural area at the fringes of Albuquerque. Due to the presence of a high groundwater table, Professor Lusk was interested in removing as much of the pollutants as possible from his residential wastewater before discharging it into the ground. One of the main problems in his particular area of Albuquerque was the lack of city sewers, with resulting contamination of groundwater and numerous individual wells from the profusion of septic tanks and leach fields.

The biological treatment system of household liquid wastes provides the basis for virtually the entire front yard landscape of this residence. Professor Lusk designed and built a system that incorporates a raised wetland cell, gravity flow into

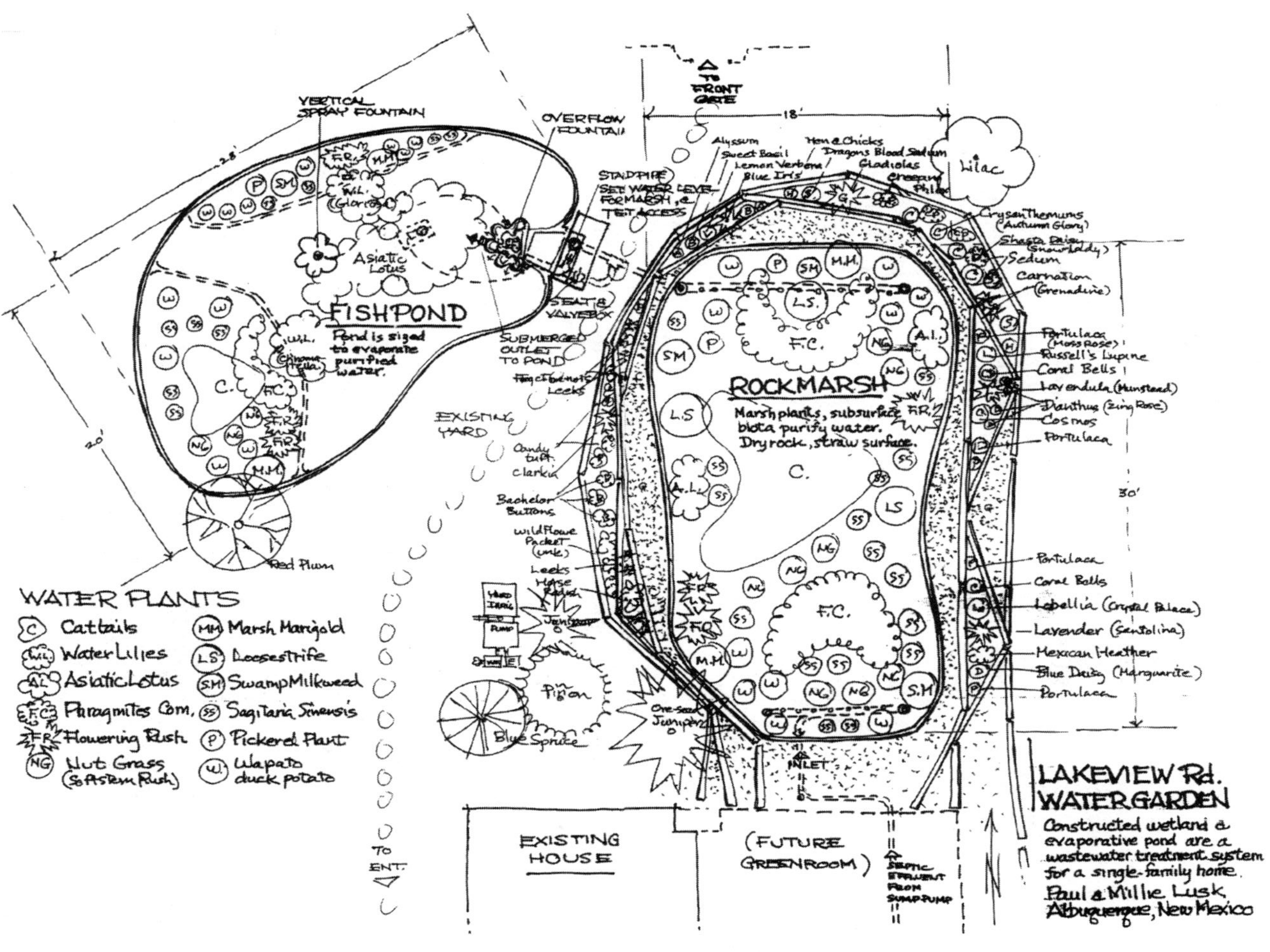

Figure 6.6 *Lusk residence, Albuquerque (Paul Lusk).*

a decorative fish pond, and a small pump recirculating system supplying a decorative fountain. Among the plantings in the wetland cell are cattail, bulrush, common reed, duck potato, loosestrife, and yellow iris. The pond is planted with Asiatic lotus, water lilies, softstem bulrush, pickerel weed, and other plants. The cost of the materials for the marsh, pond system, and lift pump was just over $2,000 in 1993. Ferro cement was utilized for the marsh and pond liners. Lusk undertook most of the labor himself; he estimates an additional $1,500 to $2,000 for labor costs. He has also designed a system that will take the biological treatment into the interior of a new greenhouse, utilizing elevated aeration tanks, gravity-fed ferrocement rock marsh, and fish pond containers. This space will also incorporate solar gain and solar-actuated venting, a cool tube to capture the evaporative cooling capacity of the exterior marsh system, and an innovative "cool tower" to assist in cooling the greenhouse in the summer. The design, based upon an experimental prototype tested in Tucson at the Environmental Research Laboratory of the University of Arizona, incorporates evaporative pads and a 7-foot-high interior water "shower" column within a 14-foot downdraft air column within the tower (Figures 6.6, 6.7, 6.8, 6.9).

FULLERTON RESIDENCE, NEW MEXICO

The Fullerton residence is located in Tesuque, a village just north of Santa Fe, New Mexico, in an area with relatively high groundwater that has become contaminated through the proliferation of septic tanks and leach fields. The Fullertons decided to install a constructed wetland system to gain better pollutant removal from their residential wastewater while at the same time providing an amenity in the form of a decorative pond. This installation, located on a hill with a spectacular view over

Figure 6.7 *Lusk residence, Albuquerque (Craig Campbell).*

Figure 6.8 *Lusk residence, Albuquerque* (*Craig Campbell*).

the piñon-covered natural landscape, provides an attractive anomaly in an otherwise arid setting. Located almost at the front door of the house, the wetland pond provides a mini-oasis that, with a small dock, almost provides an illusion of greater size. The subsurface flow portion of this system, covered with cattails, is located to one side of the pond and provides effluent sufficiently well treated to have no noticeable odor. The pond tends to become covered with duckweed during the summer, which provides additional nutrient removal while at the same time shading out the algae that would otherwise thrive (Figures 6.10, 6.11).

Figure 6.9 *Lusk residence, Albuquerque* (*Craig Campbell*).

Figure 6.10 *Fullerton residence, Tesuque, New Mexico (Craig Campbell).*

There are undoubtedly many other attractive single-family residential constructed wetlands in other areas of the country. Further exploration of the potential to integrate these systems into an overall sustainable and attractive regional landscape poses a challenge to landscape architects, and the range of potential alternatives is truly intriguing. The following chapter explores in more depth the nature of pond features associated with constructed wetland systems.

Figure 6.11 *Fullerton residence, Tesuque, New Mexico (Craig Campbell).*

7

The Pond

The highest good is like water.
Water gives life to the ten thousand things and does not strive.
It flows in places men reject and so is like the Tao.
—Lao Tsu (*Tao Te Ching,* translation by Gia-su-Seng and Jane English, 1972)

Ponds are unique entities that have the capacity to produce magical settings. Probably no single element in the landscape has the same ability as ponds to transform its surroundings, provide delight and enjoyment to both wildlife and humans, and display growth and change over the years. Ponds are living, breathing ecosystems that go through cycles from day to night and from season to season. During the past few years, there has been a profusion of books related to "water gardens" or ponds in response to increased public interest in water gardening. Long an object of attention in English gardens, the water garden is now a subject of great attraction to landscape architects, gardeners, and wildlife enthusiasts in North America. Botanic gardens in Denver, Brooklyn, New York, and many other cities have developed superb water gardens with displays of water lilies and other aquatic plants that have helped to stimulate increased interest in water gardens in many areas of the country. In response to this interest, there new companies are specializing in aquatic plants, and older firms such as Van Ness Water Gardens, Lilypons Water Gardens, and William Tricker have greatly expanded their offerings.

The focus of this volume is on environmentally friendly, low-cost, low-energy means of treating contaminated water and on the multiuse functions such sytems are capable of providing. While the subject of ornamental ponds in the landscape may seem a bit tangential to this subject, the fact that it is possible in some situations to create extremely attractive water features with the constructed wetland effluent makes it an appropriate topic to explore. There are examples in this volume that illustrate such ponds at an elementary school and a residence; and opportunities undoubtedly exist for more exploration of the feasibility of weaving similar features into other types of treatment wetland projects. In terms of sustainability, such a

Figure 7.1 *Sol y Sombra wetland treatment cells (Craig Campbell).*

system would rely primarily on gravity flow, with only a small recirculation pump to increase aeration and a biological filter. There are solar pumps that allow for a truly energy-efficient installation, and more efficient solar panels are being developed every year. From that standpoint, digression to the subject of ornamental ponds is an appropriate one, defensible within the framework of the sustainable landscape (Figures 7.1, 7.2).

Figure 7.2 *Sol y Sombra, wetland receiving pond using gravity flow (Craig Campbell).*

There are many excellent, widely available books on the subject of water gardens. Some of them emphasize the plants, some the fish, others the construction details for streams and ponds for homeowners interested in constructing their own water features. Most of these books address, to one degree or another, questions of pond biology and balance, algae, appropriate plants, and other elements of interest to water gardeners. Out of the many recent books on water gardens, two of my personal favorites, are *The Water Gardener* (1993) by Anthony Archer-Wills and *Water in the Garden* (1991) by James Allison. Most, if not all, of the information contained in these books is equally applicable to persons creating a pond utilizing potable water or those developing a pond as a final decorative "polishing" unit for a constructed wetland residential system. One of the most accessible and well-presented reference books on pond ecology is Michael Caduto's *Pond and Brook* (1985), which contains excellent descriptions and illustrations of the more common insects, crustaceans, and other inhabitants of ponds. Although the total range of microscopic bacteria, zooplankton, algae, and macrophytes (large plants) inhabiting a pond environment may vary, there are a number of typical pond species that appear to have worldwide distribution.

Ponds can be developed, of course, at any scale, ranging from a backyard landscape feature up to water bodies that are much larger and that may provide stormwater storage and treatment for large areas or wastewater polishing for major treatment facilities. Details of stormwater treatment ponds are presented in this volume in Chapter 5, and additional related information on single-family residential systems is presented in Chapter 6.

Ponds, or surface water features as opposed to subsurface flow constructed wetlands, are a necessary component of any constructed wetlands that have, as their objectives, the creation of wildlife habitat and aesthetic features in the landscape. They are also probably the most poorly understood and poorly designed elements in the landscape due, at least in part, to designers' lack of understanding of pond biology.

The interrelationships between water depth, temperature, circulation, aeration, pollutant loadings, and other critical elements often leads to ponds with unsightly maintenance problems. There can be a variety of legitimate objectives for ponds that affect the design in different ways. A pond might be designed for plants only, with no fish, and thus pose less of a challenge in regard to future overloading of fish wastes. At the other extreme would be a pond with a treated wastewater input that is also intended as a total wildlife habitat for all forms of aquatic life, including fish. The key elements in most cases are proper pond depth, proper circulation, aeration, and a correct mix of plant types.

There is often a need, particularly in arid zones, for plants that will withstand both inundation and relatively dry conditions. This requirement is particularly important for stormwater wetlands, but also for infiltration swales and basins. One of the grasses that has proven capable of withstanding wide variations in soil moisture is reed canarygrass (*Phalaris arundinacea*), which has been used to vegetate and stabilize marsh dikes due to its good root structure; its seeds are also a good wildlife food source. Another good plant for biofiltration swales and other areas exposed to varying moisture conditions, particularly in the West, is the Nebraska sedge (*Carex nebrascensis*), which is alkali tolerant, strongly rhizomatous, widely distributed, and an excellent soil stabilizer.

THE PROBLEM OF ALGAE AND POND CLARITY

Some of the inhabitants of almost every pond worldwide for at least part of the year are algae species which cause more misunderstanding than any other element of a pond ecosystem. As Michael Caduto (1985) put it, "algae is as important to a pond, lake, or stream as grass is to a field of grazing cows. Most animal life in a pond either eats algae directly or feeds on smaller animals that eat algae"(p. 62). As he points out, algae, along with other green plants and some autotropohic bacteria, are *producers.* It has been proven that many algae species exhibit maximum photosynthesis at higher light levels than most vascular plants, which show maximum photosynthesis at light levels much lower than those created by full summer sunlight. This is one of the reasons for the seasonal burst in algae growth after most growth in the higher plants has already occurred.

The next step in the hierarchy is represented by the *consumers,* or the animals and plants that consume other animals or plants. Algae are responsible for a large percentage of the total oxygen production in a pond. Photosynthesis occurs whenever energy from the sun, in *photon* units, strikes green plants such as algae. During photosynthesis—in the daylight hours—ponds absorb carbon dioxide and produce oxygen through the plants; during *respiration*—at night—oxygen is consumed and carbon dioxide given off by the plants as the organisms metabolize food molecules. This daily gas cycle occurs in both aquatic and terrestrial ecosystems. The tufted forms of algae attached to the sides of a pool or pond are generally considered beneficial and are a primary source of oxygen in a pond. The bright green, slimy form of algae is usually *Spirogyra,* which grows in strands. Algae are classified as phytoplankton and are eaten by zooplankton (small herbivores), which in turn are eaten by fish and amphibians. *Daphnia,* or water fleas, are tiny crustaceans that form large colonies and feed primarily on free-floating algae. They have the ability to clear water quite rapidly and are eaten by fish.

Although algae may pose an aesthetic problem and may be considered a pest by pond managers, these unique plants play a vital role and usually do not persist in a properly balanced pond. Their diversity of forms can be elements of great mystery to children lucky enough to be able to view them through microscopes. Even a small pond which is replenished with tap water, which contains dissolved minerals and salts, will provide nourishment for algae. In arid zones, the high evaporation rate that mandates continual addition of fresh water will stimulate excess algal growth. In general, floating leaved plants such as water lilies tend to suppress algal growth by reducing available sunlight, and submerged oxygenating plants such as *Elodea* species tend to outcompete some forms of algae for available nutrients once they are established. The difficulty is in getting such plants established in the beginning, when suspended forms of algae interfere with transmission of sufficient light to the submerged plants. Hériteau and Thomas (1994) describe in the simplest terms the biological processes of a pond, along with the importance of algae in the nitrogen cycle of a pond as follows:

Fish food, ingested, is followed by
fish waste, which releases
ammonia, a suffocating gas, which

bacteria, living under mosslike algae on the pool sides, transform into
nitrites, which other
bacteria, beneath the mossy algae, transform into
nitrates, a fertilizer, which is grabbed by the
submerged and ornamental plants, and which feeds the
algae, which greens the water, and which
fish eat in small quanitites (p. 39)

GAS AND NUTRIENT CYCLING IN PONDS

The gas and nutrient cycling processes in a pond or wetland, including the nitrogen cycle, the phosphorus cycle, and other cycles, are the result of a continuous interaction between the atmosphere, aquatic plants and other organisms, and the sun. The chemistry involved in the complex transformations of one type of nitrogen, for example, in the nitrogen cycle into other forms rely on various bacteria, some of which are denitrifying and others of which are nitrogen fixing. The same nitrogen cycle that occurs in freshwater and marine aquariums also occurs within a pond. In either a pond with fish or one receiving treated effluent from a constructed wetland, ammonia is introduced, stimulating the growth of *Nitrosomonas* bacteria, which use the ammonia as a food and transform it into nitrite. At that point, another bacteria, the *Nitrobacter,* transform the nitrite into nitrate, and *Pseudomonos* bacteria, the denitrifiers, change nitrates into nitrogen gas, thus returning nitrogen to the atmosphere. As nitrates are considered one of the most serious contaminants associated with septage waste leaching into groundwater, the pond/wetland provides a natural means of transforming much of the dangerous form of nitrogen into a benign form.

The *decomposers* that inhabit a pond or wetland environment are primarily anaerobic bacteria and some fungi, along with some aerobic bacteria that reduce nutrients introduced into a pond or wetland into basic elements available for plant growth.

The dissolved oxygen (DO) level of a pond or wetland is a measure of the actual DO measured in parts per million (ppm), which is the same as milligrams per liter (mg/L). The maximum DO that water is capable of carrying varies with temperature, with cold water containing higher levels of DO than warm water. At a temperature of 0°C, almost double the amount of oxygen can be found dissolved in the water than at 30°C. In addition, potential DO levels vary with altitude, with less potential DO at higher than at lower elevations.

DO levels are related to the photosynthesis process, as aquatic plants, including algae, produce considerable amounts of oxygen during the daytime. DO levels in ponds, therefore, are highest in late afternoon and lowest just before sunrise. If light levels during daytime hours are abnormally low for an extended period of time, respiration will exceed photosynthesis and plants in a pond will use more oxygen than they produce. Under such conditions, the aquatic plants reduce DO concentrations. Algae also contributes to this process, as does a floating mat of duckweed, which shades the water column below and lowers the photosynthesis rate of submerged species. On the other hand, the contact between the water and the atmosphere is reduced by such a mat, and the oxygen loss from the surface is attenuated. The ability to take in oxygen is also affected by pressure; plants can take in more oxygen at greater depths than they can nearer the surface of the water.

Water hyacinth, while capable of assimilating considerable amounts of nutrients, at the same time produces prodigious amounts of organic matter that are shed into the water, adding to the BOD as they decay. It has been estimated that the BOD of the daily pollution load produced in a body of water by 1 acre of water hyacinths is equal to thast of domestic sewage generated by forty people!

One floating plant that is more benign than water hyacinth is *Azolla,* a tiny aquatic fern that has a symbiotic relationship with a blue-green alga. The nitrogen-fixing alga *Anaebaena azollae* inhabits small pouches on the lower surfaces of *Azolla* fronds, providing fixed nitrogen for the fern, just as the fern provides both protection and nutrients to the alga. *Azolla* has been cultivated for centuries in Asian countries, where it is carefully managed, overwintered in special areas, and introduced as green manure in rice paddies at the proper time. As a nitrogen source for crops, it is invaluable. It also provides nitrogen to other aquatic plants upon death and decomposition. The plant is very prolific; under suitable conditions, it doubles its weight every three to five days and fixes nitrogen at a higher rate than the terrestrial legume/rhizobium association.

POND WATER pH

The relationship between water pH and aquatic plants is important but not completely understood, as pH is closely interrelated with levels of calcium, magnesium, carbonates, and other constituents of water. In general, as free carbon dioxide is removed from the water during active photosynthesis, less carbonic acid is present in the water, which causes a rise in the pH. When respiration exceeds photosynthesis, carbon dioxide is added to the water and the pH is lowered. The pH of a typical pond, therefore, is lower during the night than during the day; the fluctuation may range up to three full units between sunrise and sunset. This is typical only in a poorly buffered pond with low levels of carbonates, bicarbonates, and phosphates.

Normal rainwater has a pH of 5.6; ponds tend to develop a higher pH, which ideally should be around 6.5 to 7.5—in other words, close to neutral, which is pH 7.0. I have noticed no problems with aquatic plants or with fish such as golden orfes, koi, and native fathead minnows in my own pond with a pH of up to 8.5. A high pH is sometimes caused by too many fish or too much fish food in pond, which creates a high ammonia level in the water. However, many plants and fish are able to adapt to a particular level of acidity or alkalinity, and many suffer if a sudden change in pH is introduced through addition of chemicals.

The pH value is influenced by carbon dioxide, which can be taken in by the water directly from the air; calcium carbonate, which is insoluble in water, is used as a measure of water hardness. Water with a hardness of up to 80 mg/L is soft; water at 100–200 mg/L is medium hard; and water up to 300 mg/L is hard. The insolubility of calcium carbonate causes the precipitation of this material in a white deposit on pond edges and plants when water is hard.

Water has the interesting property of being most dense, or heaviest, when it is at a temperature of +4°C(39.2°F). Therefore, water layers that are either warmer or colder than 4°C will float above any water that is at 4°C. This characteristic, among others, contributes to the mixing and seasonal turnover of water in larger ponds and lakes.

DESIGN OF CONSTRUCTED PONDS

The advantages of incorporating a pond into a constructed wetland system, particularly a pond that is accessible to visitors, are numerous, but they have to be weighed against maintenance considerations, questions of safety and liability, and other issues that are addressed later. A well-designed pond, particular one with visible inlets or outlets that allows for some water movement, can strongly influence or temper the entire character of the larger space in which it is placed. When multiple functions such as interpretive educational features, wildlife habitat elements, and stormwater or wastewater renovation are combined in the same facility, we can experience the optimum connection between human experience and biological and ecological processes.

While there is a wide range of regional conditions and environmental parameters affecting pond design from one part of the country—or the planet—to another, there are some basic elements that have relevance anywhere. As ponds are the subject here rather than lakes, a minimum level of definition is in order. The deepest water in which emergent plants such as cattails or bulrush can grow is about 6.6 feet. The deeper a pond is, the cooler the water during the hottest months and the less algae production. Ponds less than 3 feet deep tend to heat up quickly and produce more algae during the summer.

Algae are borne on the wind by invisible spores, and will establish and grow anywhere they find warm water, light, and nutrients. There are three main types of algae, along with several other less common ones. Green algae are the most abundant form of algae to be found in ponds; they have cells with nuclei and may occur as single cells, flattened colonies, and as filaments. *Spirogyra* is a green alga that often forms dense floating blankets on the surface of ponds in spring. The various forms of blue-green algae have cells that lack nuclei, are slimy, and are the principal type causing the murky, green-brown water during early summer. In most kinds of blue-green algae, the cells stick together to form slender filaments. Some forms of these algae may float or become attached to objects and are capable of living in flowing streams when attached to rocks. Some forms of algae, such as *euglenoids, dinoflagellates,* and *diatoms,* are capable of independent swimming and might as well also be classified as one-celled animals.

There have been many successful applications of algae in *scrubbers* or filter columns that have proven the ability of specific algae species to attract metal ions and other pollutants very effectively.

Pond Liner Material

Many products on the market have been utilized for pond liners or membranes. These range from materials designed to be worked into the existing soil to render it impermeable to fabricated rubber and plastic liners installed over a prepared smooth soil bed or sand. Due to the cost of the equipment necessary to spread, mix, and tamp materials designed to be combined with existing soils, this method is normally practical only for large ponds and lakes. Various products are available, ranging from bentonite, a naturally occurring clay material that swells up to provide a good watertight seal when wet, to petroleum-based resinous polymer emulsions. The polymer emulsions, usually combined with catalytic enzymes, work by changing the forces that limit how tightly soil particles can be pushed together, such as physical shape, particle charge, and surface tension.

For the majority of small to medium-sized ponds, however, the most economical choice is a synthetic liner made of many materials. Each material typically has its own method of attaching seams, and each has a different characteristics with both advantages and limitations. Among the most common plastic type liner materials are polyethylene (PE), Hypalon (CSPE), high-density polyethylene (HDPE), polyvinyl chloride (PVC), polypropylene (PPE), butyl rubber, and ethylene propylene diene monomer polymer (EPDM rubber). While many claims are made for the benefits of each of these materials, pond liners made from either butyl or EPDM synthetic rubbers are significantly more flexible and durable than the plastic-based liners. "Pond-grade" butyl is a premium-grade synthetic rubber that is widely used in British water gardens and has the advantage of over thirty years of field testing and experience to support its continued use. Generally, when underground or underwater, butyl is more effective in resisting biological decay and will fit uneven forms. While rubber materials have lower point loads or less tear resistance than plastics, they have better resistance to blunt impacts. As possibly the best material for ponds, butyl also carries one of the highest price tags. Some butyls have EPDM added for certain characteristics, and some EPDM products have butyl added.

"Fish-grade" EPDM is one of the best materials for ponds when cost, durability, life span, and compatibility with fish are considered. EPDM is better suited than butyl for UV exposure in installations where it might be exposed to direct sunlight. It is important to understand, however, that there are many different ways to formulate EPDM; the most common use of this material has been for waterproofing roofs. A number of curing compounds and fillers used in both butyl and common EPDM can be toxic to aquatic life. EPDM manufactured for roofing membranes may contain titanium oxide; if the material is white, it is definitely not suitable for ponds. Therefore, it is essential to call for fish-grade EPDM when specifying this material for pond linings intended for display of fish and aquatic plants. Although it may be difficult to determine from the appearance of the material alone whether the material is fish grade, this information should be stamped on the box or the product should be obtained from suppliers specializing in garden pond liner materials.

Pond liner materials come in various widths, ranging from 5.5 feet for butyl to 9.5 feet for EPDM to 22 feet for HDPE. Although for larger ponds it is most common to have the liner fabricated off site, there are limitations on maximum size even when seams are factory installed. For butyl, the maximum is approximately 60 feet in width; for EPDM, liners can be fabricated to over 200 feet in width. With a special vulcanizing machine, seams can be field installed with butyl liners using exactly the same techniques utilized with prefabricated factory seams. The machine, however, is expensive, and few contractors have access to one unless they specialize in butyl pond installations.

Thickness of Liners

Thickness of liners is typically given in *mils,* each mil equaling one-thousanth of an inch or one-fortieth of a millimeter. Although the thickness of a liner doesn't affect it's aging characteristics, it can impact the ease of installation and resistance to tears and punctures. Butyl or EPDM liners of 30 mil (0.75 mm.) are suitable for small ponds of up to approximately 2,500 square feet. These liners are flexible enough to be shaped them to irregular forms and contours with little effort.

For larger ponds, or for smaller ponds with simple rectangular shapes, 40 mil (1.0 mm) or 45 mil (1.1 mm) should be utilized. A liner protection fabric manufactured from a tough synthetic nonwoven polyester, laid below a pond liner, will improve the puncture resistance of the liner and is especially important if an absolutely smooth, rock-free surface cannot be provided on which to install the liner. These liners are approximately one-eighth of an inch thick and very flexible; they do not deteriorate with age under a wide range of soil pH conditions and are nonbiodegradable. The liner fabric also protects against punctures from objects above the liner. It is good practice to install a protection fabric above as well as below the liner when the liner is to be covered with soil, gravel, or rock.

When large ponds are part of a project, it may be necessary to utilize lower-cost materials such as PVC (fish grade if fish are to be introduced into the pond) or HDPE. Standards within the liner industry can be quite confusing, as in many cases the specifications differ from one product to another. Costs for EPDM range from around 50 cents per square foot for 45 mil to over $1 per square foot for 40 mil butyl rubber.

Penetrations of pond liners to accommodate inflow pipes, recirculation system piping, or drains must be done with the greatest care. In general, if all necessary inflow and outflow can be managed with surface flow over the top of the liner, penetration should be avoided. If it is unavoidable, premolded pipe boots made of various materials are available to provide the most fail-safe method of liner penetration. The boots incorporate a short section of pipe or a conical section that can be cut off at the proper length to provide the required diameter for connection to a pipe. A collar of butyl, EPDM, PVC, and a variety of other liner materials securely attached to the pipe or cone allows for a solvent connection to the liner in locations where penetration is required. The precise depth and location of such penetrations must be carefully planned in advance to avoid relocation of a poorly placed boot. Boots made of rubber material for butyl liners are generally field applied with cold-seaming tapes.

Edge Design

The single most difficult element in the design of a successful pond is the design and detailing of the pond edge. Too often, a poor design results in exposure of the liner around the pond edge, which can accelerate deterioration with some liner materials and which, in any case, is an eyesore. There are a number of ways to place boulders and small rocks around an edge to create a natural-looking edge while at the same time protecting the stability of the edge and offering spaces for plants. For the most successful display of ornamental aquatic plants that prefer shallow water, a planting ledge must be provided that also presents a safety area between the pond edge and the deeper water. Rock selection and placement, is truly an art, and demands patience along with sensitive and sometimes repeated efforts to achieve just the right exposure of rock. Rock that has been naturally rounded by exposure in a river bed can be an excellent "accent" rock, but more angular rock is easier to place around an edge. The flexibility of rubber-type liners offers much more protection against punctures than plastic liners when rockwork is accomplished over them. If a pond is being built by an individual, as opposed to a contractor, then the rock should be kept to a size that can be handled by one person—usually a rock no more than 6 to 8 inches thick by 8 to 12 inches wide and up to 24 inches long. In much of the West, weathered sandstone and granite rock are commonly available, many with colorful lichens attached to their surfaces (Figures 7.3, 7.4, 7.5, 7.6).

The largest rocks used in many areas in the West for pond enhancement are typically rounded granite or sandstone boulders, some of which feature attractive and colorful lichens. Several rock and pond supply or construction firms have obtained the right to remove naturally weathered surface rock of up to 10 tons or more, many with spectacular sculpted water channels providing singular works of natural art that can be integrated into water feature installations (Figure 7.7).

There are many different ways of incorporating rock into pond or channel edges. However, in all cases, rocks should be placed only on liners, with a protection fabric below the liner to improve puncture resistance. In addition, it is helpful to install another layer of liner protection fabric over the liner, under the rock, to protect the liner when rocks must be nudged into precise loctions (Figure 7.8).).

Pond Soil

In general, any soil with a high clay content will be suitable either for pond bottom or for planting ledges. If any soil is available from existing ponds or wetlands that

Figure 7.3 *Typical detail with rock edge extending into the water (Resource Conservation Technology).*

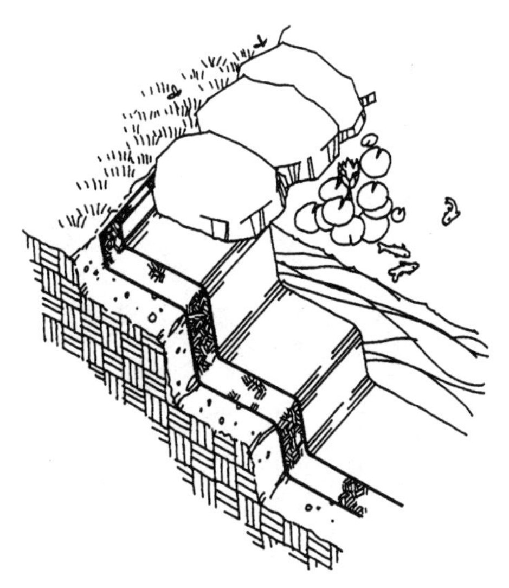

Figure 7.4 *Typical detail with rock edge above the water level* (*Resource Conservation Technology*).

are being dredged, it may provide a balance of microorganisms from the beginning, although it will not take long for these microorganisms to become established in any new pond on their own. One mixture widely used in Europe is seasoned compost with one-quarter sand and some peat. One good material almost universally available is *crusher fines* consisting of a mixture of sand and stone up to about 1/4 inch in size—essentially the leftover material at gravel pits after screening for larger rock. This material can be used as a pond bottom material over liners in order to minimize any cloudiness caused by fine particles going into suspension.

Figure 7.5 *Typical detail for pond divider and bog garden* (*Resource Conservation Technology*).

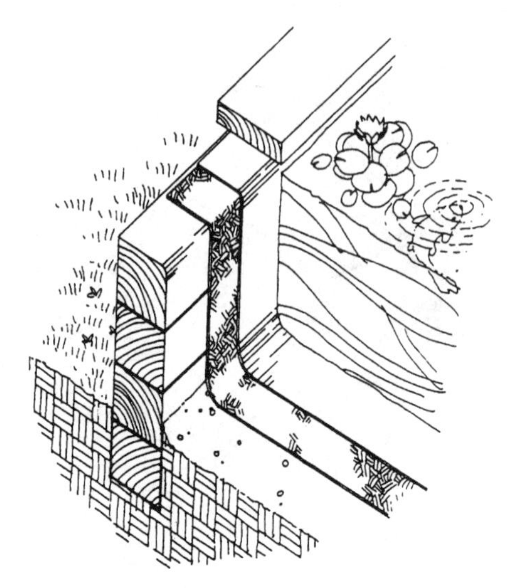

Figure 7.6 *Typical detail for timber edge (Resource Conservation Technology).*

The use of a 1- to 2-inch layer of sand or crusher fines over a soil with clay material will produce the same results.

Biological Pond Filters

The increasing interest in water gardens internationally has spawned a raft of books on the subject, as well as more widespread experimentation with simple means of

Figure 7.7 *Garden pond with sandstone edge, Campbell residence. Plants include pickerel weed, narrow leaf cattail, yellow iris, and water lilies (Craig Campbell).*

Figure 7.8 *Los Padillas Elementary School pond with granite boulders (Craig Campbell).*

keeping pond water as clear as possible. The development of purely aesthetic backyard ponds is addressed in depth in many books on the subject and is beyond the scope of this volume, but it is worthwhile to describe one of the more interesting offshoots of the water gardening industry.

Biological filters are now offered by many water garden supply companies as well as being fabricated locally by many suppliers of aquatic plants. No true filtration occurs in biological filters in that no particles are screened and removed from the water. Instead, almost any object left in the water long enough will become a biological filter in terms of providing an environment for beneficial nitrifying bacteria to convert ammonia into nitrites and nitrates. A typical sewage treatment plant trickling filter, which sprays water over a layer of water in a tank, is essentially a biological filter. At a greatly reduced scale for pond applications, a biological filter provides an optimum environment for the greatest number of bacteria within the smallest "processing" space to thrive and thus metabolize the ammonia present in a pond.

Nitrifying bacteria will colonize most submerged surfaces in a pond, and will spread and multiply as long as enough ammonia and oxygen are present. By concentrating the potential surface area, such as in a gravel or special plastic medium, and ensuring an oxygenated environment, a biological filter can accomplish much more efficient cleansing of a small body of water compared to natural cleansing. Material on a pond bottom soon becomes clogged with both living and dead biomass, which tends to create anaerobic conditions, favoring bacteria that thrive in the absence of oxygen but that are not efficient nitrifiers. Biological filters need to be designed to operate without clogging and are generally of two types: internal, constructed within the pond, and external with a flow and vetura pipe. The most common type pumps a certain volume of water from a pond continuously, usually upflow, through a gravel or other medium. The biological filters operating with a water pump unfortunately require pumps to run continuously and may fail to perform properly if not in operation all the time. The organisms may start to suffer after only a half hour of interrupted flow.

Although the need for electricity on a continuous basis greatly reduces the value of such systems in the overall context of energy conservation and natural systems, there are solar-operated aeration and pumping systems that may be combined with a biological filter to either minimize or entirely eliminate the need for additional electrical power. Storage batteries are required to provide continuous operation. Currently on the market are solar "power stations" for aeration and pumping systems ranging from 50- watt units up to 1,300-watt units at costs ranging from $300 to $1,300. There are also wind aerators available in various models designed for wind speeds from 1 to 15 mph. Such systems can be combined with solar systems to provide ice removal in ponds or small lakes, along with increased aeration.

Another filter type, patented by Aquacube, operates by employing an energy-efficient regenerative air blower that requires no submerged electrical wires and can substantially reduce algae. These filters use air blowers that create a flow of low-pressure air that enters submerged diffusers at the base of the filter unit. The air is diffused up through a substrate, creating an upward flow of air, and through water, creating an optimal aerobic environment for the colonization of beneficial bacteria. These units are capable of circulating more than 400 gallons per minutes per surface square foot of the filter unit; a one-half-horsepower unit can continuously circulate over 4,000 gallons per minute of biofiltered water. These units are often combined with a bead filter in koi ponds to create crystal clear water.

It takes three to eight weeks for beneficial filter organisms to establish themselves in the medium when a biological filter is initially installed. This process can be somewhat speeded up by "seeding" the medium with bacterial cultures that are available from pond supply companies. The main purpose of these filters is to satisfy the need of pond owners or managers, particularly koi enthusiasts, who prefer the appearance of crystal clear water to that occupied by naturally occurring algal populations during part of the year. Others with shallow ponds, who do not expect them to be crystal clear, preferring the more mysterious aspect of a somewhat murky water, can live with seasonal "green" water or can even add a black vegetable dye to a pond to increase its reflectiveness and accentuate the plantings, especially floating plants such as water lilies. This also tends to obscure the planting pots in shallow ponds that otherwise may be objectionable.

There is, however, at least one other type of design for a biological filter that does not rely on continuous operation of a pump, although it is not as effective as those previously described. This method incorporates a *filter pool* that is located at an elevation above the main pond and receives all recirculated water. The design of this type of filter pool includes a gravel filter medium at least 12 inches in depth, with 4-inch perforated pipe spaced 2 to 3 feet apart below the gravel. All recirculated water must pass through the filter medium, after which it flows up an over a spillway into the main pond. The advantage of this method is that the medium retains water at all times and thus is not prone to bacteria dieoff if a recirculation pump is not operated for a few days. Additional treatment can be obtained by installing water hyacinths in the filter pond to assist in removing excess nutrients from the water. This type of system will usually operate for year with no servicing, but if filter pipes become clogged, water can be pumped into the chamber to force it through the filter medium and thus unclog the pipes. The gravel media in these systems undoubtedly have to be cleaned out periodically as well.

RECIRCULATION PUMPS

For small ponds up to 1,000 to 2,000 square feet in surface area, submersible pumps are the most efficient way to recirculate the water through a fountain display, waterfall, or stream. Most companies specializing in water garden plants and fish also handle pumps; they can also be obtained from most building supply stores. Submersible pumps are available in sizes ranging from very small, low-voltage types to fractional horsepower 110-volt pumps capable of moving 1,200 gallons per hour or more. It may difficult for a person to determiune the right amount of water for a constructed streambed and waterfall without testing several pumps, but a 1,200-gallon-per-hour, one-sixth-horsepower pump will create an attractive flow of water in a small streambed 18 inches in width. It is important to consider the long-range costs of electricity when installing a pump as large as one-half horsepower, particularly if the pump will operate more than an hour or two a day. A one-half-horsepower pump is capable of pumping up to 4,000 gallons per hour, but uses 1,100 watts, as opposed to a 1,200-gallon-per-hour pump using 380 watts. Pump selection must also take into account the vertical head, or difference in elevation, between the location of the pump and the highest outlet that is receiving the recirculated water. Most submersible pumps have no problem lifting water to 10 feet or higher. There are a number of manufacturers of submersible pumps designed primarily for garden ponds; Little Giant and TetraPond are just two of those that are widely available.

Solar-powered pumps were mentioned at the beginning of this chapter. They are more appropriate within the framework of a sustainable landscape than a standard pump. As collector technology improves, larger pumps will probably be available at a more reasonable cost than at present. They are currently available for up to 20 gallons per minute.

FISH IN PONDS

In most parts of the country, mosquito fish (*Gambusia affinis*) have proven to be very useful in controlling mosquito larvae. They are silvery-gray, grow to about 2 1/2 inches long, and reproduce quickly. They are considered a very cost-effective means of controlling mosquitos in irrigation ditches and ponds in the Albuquerque, New Mexico, area, and the county health department provides them free of charge to all residents. An interesting characteristic of mosquito fish is their ability to adapt their coloring, becoming darker or lighter almost instantly when needed to blend in to the surroundings and provide protective coloration. They have been successfully stocked and overwintered in climates as cold as those of Michigan and Ohio.

While it may be desirable to stock only mosquito fish or native fish in larger ponds, in smaller ornamental ponds the sight of more colorful introductions is worth the extra attention they may require. It must be recognized, however, that the concentration of un-ionized ammonia may be too high in a pond receiving effluent from a constructed wetland for many fish to survive. The limit for mosquito fish is around 1.1 mg/L; and for carp (including koi) it is less than half that concentration. Therefore, additional special methods may have to be employed to ensure ammonia reduction in effluent prior to discharging directly to a pond if fish are desirable.

There are two extremes represented by climate and water temperature, each with its own limitations and possibilities: very cold waters and tropical waters. The options for these situations are well described in other volumes and are not addressed here. The majority of ponds in this country probably are between the two extremes and may experience ice in the winter months. There are a number of fish that can thrive even when winters produce a thick cover of ice over a pond. Among the fish species appropriate for larger ponds are fathead minnows, bluegill, and green sunfish.

Fathead minnows (*Pimephales promelas*) occur in nature or have become established virtually throughout the United States and provide a major source of bait minnows for fishermen. They can tolerate very small ponds and extremes in pH and turbidity; they are also capable of tolerating salinity levels up to at least 10 percent. However, while fathead minnows can survive in small ponds, they also are capable of rapid reproduction and can overwhelm more colorful species in a very short time.

Bluegill (*Lepomis microchirus*) and other related sunfish are often stocked in farm ponds or bass ponds, where they reproduce quickly and provide the major food source. Bluegill prefer clear water with aquatic plants and bottom weeds, and are suitable only for larger ponds.

Green sunfish (*Lepomis cyanellus*) are more tolerant of turbidity than bluegill and can even survive in temporary ponds and swamps. Like all sunfish/bluegill, they are suitable only for larger ponds due to their rapid reproduction rate.

Among the fish species that are more suitable for small ponds, due to their attractive coloration and relatively small size, are various goldfish species, golden orfes, and koi.

Goldfish (*Carassius auratus*) have a long history, dating back to at least the seventeenth century in China. They are ideal for small garden ponds and are hardy, remaining active in summer when water temperatures reach as much as 95°F (35°C) and withstanding oxygen levels as low as 25 percent. There is a large number of goldfish types. One of the most beautiful is the shubunkin, which has transparent scales and a color range from gray-blue to bright crimson with black spots. It thrives in warm water and breeds readily even in very small ponds.

Golden orfes (*Leuciscus idus*) are another good choice for small or large ponds. Orfes are surface feeders that devour midges and mosquitoes. They are a light carrot orange in color and dart about quickly, often in schools. Orfe populations can double in size each year for the first several years but seldom breed in small ponds, preferring larger bodies of water.

Koi are simply a fancy colored carp that originated in Japan, where countless variations are bred and given names, some fetching extremely high prices. The *Showa sanke,* for example, sports red and white markings on black. They can become quite tame and can be trained to eat out of ones' hand. They may be raised in small ponds until they grow to a size that is incompatible with the pond environment. In most cases, however, koi will not grow as fast or as large in small ponds as in large ones; they do have the potential to grow quite large and to root around in plants.

POND MAINTENANCE

Ornamental ponds are not generally self-maintaining; as with any other element in the landscape, an intelligent understanding of the ecology and biological demands of the system is essential to good management. The indiscriminate reliance in our culture on a quick chemical fix for pest problems, for antiseptic cleaning, and for myriad other daily activities produces biological effects downstream but normally not visible or obvious to the user. By contrast, when many commonly used chemicals are placed in a septic tank, wetland, or pond, the effects may be immediate.

The use of insecticides and herbicides over the past thirty to forty years has resulted in serious damage to the natural environment and to wildlife while enhancing the development of resistance within both insect and plant populations. Many commonly used herbicides and insecticides may persist for many years in certain environments and are especially deadly to aquatic organisms. The movement toward *integrated pest management* methods that emphasize biological control of pests, along with cultural practices that minimize the opportunities for proliferation of pest species, is long overdue and is now gaining wide acceptance. It has been clearly demonstrated that even when many insecticides are applied at the authorized rates, they tend to increase in concentration in each succeeding level of the food chain, a process known as *biomagnification.*

That is why the Hibernia development in Highland Park, Illinois, has covenants restricting the use of certain lawn care products, solvents, herbicides, deicing materials, and other chemicals that could impact their wetlands. They have developed an educational program designed to promote self-policing, which has not deterred sales in the slightest. Bellevue, Washington, also actively discourages the use of pesticides and herbicides due to the potentially damaging environmental effects on groundwater and stream quality in the area.

Certain forms of algae can present a problem for pond owners. The range of types and microscopic forms of algae is remarkable, with some being associated with clean water and others with polluted water. For control of algae in garden ponds, there are a number of commercial products, in addition to biological filters, which do not harm either aquatic plants or fish. TetraPond, one of the major suppliers of fish pond supplies, markets a "water clarifier" named Aqua Rem that essentially binds together suspended algae into clumps that sink to the bottom. These pond additives are fairly innocuous if algae problems are not severe or if the pond bottom is vacuumed every year; however, if a pond has a severe algae problem, the large amount of material that falls to the bottom following treatment will decompose and lower the oxygen level, which may cause fish to gasp at the water's surface. Under these circumstances, the decomposing material on the bottom should be pumped out and the water oxygenated with a fountain, "air stone," or pond air pump. Some pond supply companies sell what they call *filtration bacteria* in both liquid and dry form. While these products may be of use in "spiking" a biofiltration filter, they are of dubious value for reducing algae when dumped directly into a pond. Tadpoles are algae consumers, as are snails; installing tadpoles, either collected or purchased from an aquatic supply company, may help control algae problems in some ponds.

As mentioned previously, algae can be minimized by installing oxygenating plants to compete with algae for nutrients, as well as water lilies and floating plants to shade the water. In addition, fertilizers should definitely not be allowed to run off from lawns or ornamental plantings into the water, and fish should not be overfed.

Larger ponds used for wastewater or stormwater treatment and storage are also often employed as storage reservoirs for irrigation water for golf courses, athletic fields, and parks. Algae buildup can present problems in these ponds, clogging up irrigation valves and nozzles. In addition to the controls already mentioned, it is often desirable to install a filter specially designed to remove algae. One self-cleaning filter that appears to function with fewer problems than sand filters is made by Filtomat and operates by employing coarse and fine filters to screen out debris and an automatic backwash cycle that is set off by a pressure differential created by debris buildup on the stainless steel screen. In summer, when algae are plentiful, the filter backwashes about every ten minutes; in winter, when less algae are present, the filter may backwash only once per hour. The flow is not interrupted during the backwash cycle.

VISIBILITY, ACCESSIBILITY, LIABILITY

There are, unfortunately, few outstanding examples of the integration of constructed wetlands for either stormwater or wastewater treatment into an overall landscape designed for human interaction, education, recreation, and visual enjoyment. The one example that has received the most national publicity is the Arcata, California, system designed by Robert Gearheart. Others that are detailed in this book are the Los Padillas Elementary School outdoor classroom and constructed wetlands in the south valley of Albuquerque, the Greenwood Urban Wetland and the Orlando Easterly Wetlands.

The questions that immediately come to the fore whenever the concept of a constructed wetland for wastewater is proposed for public accessibility or within a built-up urban area are usually the following:

1. Questions related to the safety of the water if publicly accessible ponds are involved; these questions concern pathogens, fecal coliform, and so on.
2. Questions related to potential odor and mosquito problems.
3. Questions, in some areas, related to liability and access control for ponds.

The safety questions are reasonably well addressed through experience with existing facilities that undergo regular water quality sampling. It has been demonstrated that fecal coliform is normally within acceptable levels in the effluent from a properly designed and properly maintained constructed wetland; viruses have never been detected in these systems. In addition, state regulatory agencies generally require that any wastewater effluent be disinfected through a process of chlorination, UV treatment, or other approved means prior to discharge into an open pond accessible to the public.

Concerning questions related to odor, almost all ponds receiving effluent discharged from a constructed wetland have either no odor or an extremely faint odor at certain times of the year. The precise interaction and activity of various microbes plays a major role in assisting the process of wetland assimilation of pollutants. The usual source of odor is hydrogen sulfide, which is produced only within an extremely reduced environment, that is, an aquatic environment with little or no regular input and outflow and an entirely anoxic substrate. These conditions are not normally present in a constructed wetland that is continually receiving input and is planted with the proper vegetation capable of transporting oxygen to the root zone.

Mosquitos can be a problem but as previously mentioned, they can be controlled by introducing mosquito fish (*Gambusia*) that feed on mosquito larvae. These fish are considered such an effective control that the county health department in the Albuquerque metropolitan area provides thousands of them each year at no charge to residents to install in their irrigation ditches and ponds. Other natural controls for mosquitos are *Bacillus sphaericus,* sunfish, and bluegills. Many pond-loving invertebrates, such as dragonflies, also help control mosquito larvae. Some species of mosquito, however, attach themselves as larvae to the submerged stems of emergent aquatic plants and tap into their lacunae, or hollow tubes, to obtain their oxygen. In general, the use of multiple methods of control is favored by most pond managers. Michelle Girts, a biologist with the firm CH2MHill in Portland, Oregon, reports that the installation of bat houses and swallow nesting boxes has had good results in some areas, with the swallows assisting in mosquito control during the daytime and the bats at night.

Liability Issues

The liability issue is possibly the most difficult one to address; no clear-cut ordinances or regulations usually exist to cover ponds as opposed to swimming pools. Opinions and regulations vary widely from one jurisdiction and state to another. The general consensus seems to be that in a locale where natural ponds are common, there is no particular requirement or need to fence off newly constructed ponds, whether they are designed for stormwater treatment or for other purposes. This assumes that a reasonable safety factor is designed into the edges of the ponds, including a planting or safety shelf and a slope no greater than 3:1.

In an innovative development in Highland Park, Illinois, that incorporates both a stream and ponds as part of a wildlife habitat and stormwater conveyance and detention system, the ponds and a central lake feature an underwater safety ledge ranging from 2 to 4 feet in depth. The assumption in these cases is that there is no increase in liability beyond that which already exists, based upon the assumption that there is already a well-developed public awareness of the danger of water bodies in general.

In the Puget Sound region, the recommendations of the State Department of Ecology call for side slopes to be no steeper than 3:1 around the perimeter (as opposed to any islands), with dangerous outlet structures protected by an enclosure.

In addition, the state recommends warning signs for deep water and potential health risks; the signs must be placed so that at least one of them is visible and legible from all adjacent streets, sidewalks, or paths. When the pond surface exceeds 20,000 square feet, it is recommended that a safety bench with a width of 5 feet be provided around the basin, with a maximum depth of 1 foot during nonstorm periods. It is further recommended that this bench be planted with emergent vegetation such as cattails to inhibit entry. Fences are required only when the side slopes are vertical walls or when homeowners associations or local governments require them.

In some jurisdictions, particularly in arid areas where natural water bodies are uncommon, there seems to be greater concern about the potential liability exposure of having any water body over 18 inches in depth that is not fenced off. When such features are included in areas that can easily be reached by pedestrians within an urban context, the installation of fencing and controlled access gates is probably the best solution.

The general consensus seems to be that in areas where natural ponds and wetlands are common, there is less concern over liability issues related to the development of constructed wetlands and stormwater ponds than in arid regions or other areas with few naturally occurring water bodies. In addition, it is generally conceded that rapidly moving water in ditches, channels, and rivers seems to be more attractive than still water and thus represents more of a hazard.

Indeed, by far the greatest number of incidents occur in desert southwestern cities within concrete trapezoidal drainage channels and in large irrigation ditches; the ugly engineered concrete channels are a particularly intrusive and malevolent feature of the urban landscape. It is interesting to note that the smooth bottoms and sides of drainage channels such as these are dangerous precisely because of the nature of the design and of the materials: they are intended to convey water at high velocity over smooth surfaces that offer no opportunity for individuals who may be swept away to grab on to a handhold. Drainage conveyances such as these represent the best thinking of the period in which they were built, between the 1920s and the 1960s, but they could have been designed and constructed in a more natural manner, at less cost and with greater benefits in terms of visual effect and slowing velocity. Retrofitting such channels after the fact is costly but not impossible.

The *Stormwater Management Manual* for the Puget Sound Basin, mentioned in Chapter 5, summarizes the potential problems posed by ponds in urban areas as follows:

> The presence of wet ponds and marshes in established urban areas is perceived by many people to be undesirable. They are often thought of as mud holes where mosquitoes and other insects breed. If the wet pond has a shallow marsh established . . . , the pond can become a welcomed addition to an urban community. Constructed fresh water marshes can provide miniature wildlife refuges, and while insect populations are increased, insect predators also increase, often reducing the problem to a tolerable level.

The manual goes on to state that if there is sufficient concern, homeowners associations may want to drain ponds during late spring and summer; it is imperative that the vegetation in the shallow marsh areas not be allowed to die off during the draindown period to avoid impacting the pollutant removal effectiveness of the system.

Ponds can be an extremely important element in any landscape, whether within a private or a public realm. The important thing to recognize, however, is that these features require care and ongoing maintenance in order to look their best and to function as intended. In the absence of this commitment to a little extra attention, careful consideration should be given to the feasibility or appropriateness of a pond feature for any project.

8

Wildlife Considerations and Management

> Civilization is a state of mutual and interdependent cooperation between human animals, other animals, plants, and soils, which may be disrupted at any moment by the failure of any of them.
>
> Aldo Leopold (*Game Management,* 1933)

Constructed wetlands, regardless of size or locale or objectives, will ultimately provide an attractive habitat for some forms of wildlife that may or may not have been present on a particular site prior to installation of the wetlands. There are a number of constructed wetlands designed primarily for wastewater treatment and polishing that have inadvertently become some of the richest waterfowl breeding habitats in the regions in which they are located, and others that were designed from the beginning with the creation of wildlife habitat as an objective.

URBAN WILDLIFE

It is only fair to point out that within any given public body, opinions will probably be somewhat divided regarding the desirability of attracting particular wildlife to an area, especially near residential development. Cities such as Boulder, Colorado, have an extremely serious deer problem due to overpopulation; others have problems with too many raccoons, skunks, possums, or—in the West—prairie dogs. (Figure 8.1). Probably the greatest problem posed by wildlife, ironically as a result of the successful management program for restoration, is associated with Canadian geese, which have found attractive year-round habitats in areas formerly utilized only seasonally. Canadian geese have found favorable wintering grounds in grassy parks and campus settings from Seattle to Boston. At least several hundred thousand of them have become urban dwellers, posing serious problems due to the sheer quantity of their wastes in Seattle, Boston, Toronto, Cleveland, Chicago, Wilmington, Denver, Indianapolis, Nashville, and Minneapolis. In addition to the

Figure 8.1 *Sign in a downtown Boulder, Colorado, park (Craig Campbell).*

waste problem, geese love to feed on grass, and are therefore attracted to areas ranging from backyards to parks where they can congregate and feed.

Bears and coyotes have become urban problems, both in cities like Albuquerque and in small communities such as Los Alamos, New Mexico, which had 150 reports of nuisance bears in just one recent year. Muskrats (and nutria in certain parts of the country) can pose a serious problem within a constructed wetlands due to their propensity to burrow into dikes and even chew through liners. Muskrats reproduce rapidly. Their resulting population density, through heavy feeding on cattails and other aquatic plants, can decimate large areas of a wetland. Farmers, in particular, have a legitimate complaint related to flocks of "granivorous" birds such as ducks, blackbirds, and others attracted by nearby wetlands. Crops such as corn, rice, wheat, barley, and sunflowers all attract birds to feed.

There is growing evidence, however, that people living in most metropolitan areas have a genuine interest in wildlife and in opportunities to view wildlife that outweighs any perceived pest problems. A 1985 national survey of Americans (U.S. Fish and Wildlife Service and U.S. Department of Commerce, Bureau of the Census) estimated that 65 percent of the adult population enjoys seeing or hearing wildlife while pursuing other activities at home. It was estimated that 58 percent of Americans sixteen years of age and over maintained an active interest in wildlife around the home through activities such as observing, photographing, and feeding wildlife or maintaining natural areas for the benefit of wildlife. A survey of Canadians (Fillon, 1983) reported similar results. In a survey of residents of Kansas City, Springfield, and St. Louis, Missouri, 93 percent of the respondents described the wildlife around their homes as "enjoyable" rather than as "pests," and only 13 percent reported that they had wildlife-related problems. Surveys in some other cities indicated that residents wished the city to do more to encourage wildlife conservation. A survey of the residents of Columbia, Maryland, indicated that the residents preferred future stormwater ponds to be managed for fish and wildlife in addition to flood and sediment control; they wanted lakes and ponds to be developed for waterfowl, mammals, and other wildlife rather than for public fishing. They most preferred waterfowl, songbirds, shore and marsh birds, and frogs and turtles; they most disliked raccoons, muskrat, and snakes. This survey led to a decision in Columbia to establish a waterfowl protection zone designed to protect swans and other waterfowl from fishing hooks and tackle. A swan research program was established, and efforts are underway to restore the trumpeter swan by encouraging breeding on three lakes.

One researcher (Schicker, 1987) discovered, in a survey of children aged six to 10, that 50 percent of all their outdoor activities directly involved wildlife (e.g., observing, collecting) and that children, unlike adults, were most interested in "creepy-crawly" varieties such as amphibians, reptiles, and insects. These were mentioned, looked for, and collected more than all other forms of wildlife, which tends to support the importance of riparian and wetland habitats in our environment.

Research has indicated that contact with plants and animals is therapeutic and enriches the lives of elderly people and others requiring specialized care. The settings most likely to provide habitat for a variety of wildlife—roadsides, parks, cemeteries, school yards, and other open spaces—have not traditionally been managed for wildlife; their primary purposes have been in other areas. There have been movements within some cities, such as Portland, Oregon, to develop wildlife habitats within urban areas; and Albuquerque, New Mexico, manages a very successful wildlife sanctuary at the Rio Grande Nature Center that seasonally attracts sandhill cranes, geese, and many songbirds (Figure 8.2). The Chattanooga Nature Center employs a constructed wetland treatment system for restroom wastes, although unfortunately the system is not interpreted for visitors.

There is even a significant movement within the golf course development industry to incorporate a wider variety of vegetation to create wildlife habitat, particularly for birds, within the design of new environmentally sensitive courses. This

Figure 8.2 *Rio Grande Nature Center, Albuquerque, New Mexico (Craig Campbell).*

is undoubtedly due, at least in part, to the desire to create a more favorable image for golf courses in some areas of the country where they are not particularly viewed as beneficial, but it also involves reaching out to conservation organizations such as the Audubon Society in a cooperative manner not experienced in the past. Any move to reduce reliance on pesticides and herbicides for golf course maintenance will ultimately have beneficial effects on groundwater, streams, and wildlife habitats.

ECOLOGICALLY BASED PLANNING AND WILDLIFE

One of the more urgent needs in our rapidly expanding metropolitan areas, if not equally in our small to mid-sized cities also facing growth, is the establishment of urban wildlife reserves and corridors. The responsibility of humans for the landscape and all of its organisms has normally been neglected in overall planning and zoning structures, and maintenance of biological diversity "in our own backyards" has not given proper consideration. As one observer put it,

> the condition of our urban centers can be seen as indicators of the impact of human populations on the global environment; and the degree to which biological diversity is either encouraged or diminished in urban areas is intrinsically connected to our commitment to protection of eoosystems on a worldwide basis. In other words, we need to concentrate more seriously on nurturing stewardship on a local as well as global level.

Although many cities are blessed with admirable park systems developed in the traditional manner, with large areas of lawns interspersed with shade trees, this particular landscape structure supports much less diversity of wildlife than do landscapes with a fuller, more natural mixture of deciduous and evergreen species of different ages and multiple layers of vegetation. Even dead trees and snags help provide diversity of habitat. These characteristics, which are desirable for wildlife, of course, are also those that many public officials and park managers find problematic for humans due to the cover such landscapes provide for criminal activity and the difficulty of policing areas with vegetative screening. This is a question with no easy answer, but the obvious one that can be applied in many areas is to clearly provide and identify a hierarchy of park and wildlife facilities, each of which has characteristics and limitations that tend to determine appropriate locations.

Management of public open space for wildlife habitat is not currently a high priority for most urban park and recreation departments, and more knowledge and sensitivity in this area needs to be brought into city government. Wildlife biologists, who are highly motivated professionals trained to deal with human/wildlife interactions, have traditionally had few job opportunities outside the federal government and some state agencies. This type of knowledge and expertise should be brought down to the level of local government, where wildlife biologists would provide insights into planning and management decisions in close coordination with landscape architects, planners, engineers, and park managers. Only by integrating this perspective into the overall system of developing and managing our urban areas will we see a broader, more environmentally holistic pattern of values and objectives underlying our planning efforts.

There are several examples of cities and counties that have integrated wildlife habitat considerations into their long-range planning efforts. One of the most notable is Fort Collins, Colorado, a city of approximately 100,000 located 60 miles north of Denver. Fort Collins is certified as an urban wildlife sanctuary by the National Institute for Urban Wildlife due to the city's formal recognition of the value of wildlife habitat within the city, as established by resolution in 1987. The City Department of Natural Resources employs wildlife biologists and other specialists who interact with planners and others responsible for overseeing development and protection of all areas within the city. The department sponsors educational programs such as the "Backyard Wildlife Habitat" program, which provides information to residents on attracting wildlife. In addition, it has developed criteria for building development based upon evaluation of wildlife habitat. When schoolyards or parks are enhanced with ponds, feeding stations, and appropriate plantings, they are capable of attracting a variety of songbirds and other wildlife while at the same time improving the environment.

In some parts of the country, homeowners who prefer to develop their lots with the objective of attracting birds and other wildlife often come into conflict with local ordinances related to weed control and grass height; thankfully, there is much more recognition at the national level of the environmental benefits of native vegetation and wildflower meadows as opposed to maintained, fertilized, mowed turfgrasses. In communities where they have been challenged, many of the old weed and grass ordinances have been struck down by courts. In Little Rock, Arkansas, a judge dismissed a citation against a local resident because his "overgrown" yard was legally certified as a "backyard wildlife area" by the Arkansas Fish and Game Commission.

The National Institute for Urban Wildlife (NIUW) is actively engaged in certifying a network of urban wildlife sanctuaries on private and public lands across the nation and makes available information, some of it oriented toward teachers, on conservation and enhancementof urban wildlife habitat. In addition to the NIUW, there are several other wildlife organizations that support the development of urban and backyard wildlife habitats.

Another organization concerned with wildlife habitat is the Wildlife Habitat Enhancement Council, which sponsors workshops and conferences and attempts to develop common goals for enhancement of wildlife habitat by working with industry, environmentalists, and regulators. The Council tends to focus on corporate lands and has certified 185 programs around the country, as well as in Canada, Spain, and Australia. This organization has a special interest in habitat enhancement programs on degraded lands such as mine sites, parking lots, and landfills, which otherwise tend to be overlooked for wildlife habitat restoration.

Boulder County, Colorado, also has integrated wildlife habitat maintenance into their planning process. By identifying Environmental Conservation Areas (ECAs) along with areas of critical plant habitats and species, the county has established one of the most protective planning processes in the country related to preserving both wildlife habitat and connective corridors. Through an intensive research and review process involving biologists, botanists, ecologists, private conservation organizations, and government agencies, a planning and review process ranging from long-range planning down to the building permit level, designed to preserve the county's richest biological sites, was established, along with appropriate buffers and corridors for wildlife habitat and movement. With a large wilderness area along its western border, Boulder County contains an impressive variety of natural habitats including grasslands, riparian zones, forests, and even alpine tundra. With its commitment to preserving biodiversity in the face of major urban development pressures, Boulder County has established itself as one of the national leaders in the area of ecologically based planning.

Another city tht has integrated wildlife habitat considerations into the planning and development process is Tucson, Arizona, which has experienced phenomenal growth over the past two or three decades. A statewide wildlife habitat evaluation was undertaken by the Arizona Game and Fish Department (AGFD) in 1979, and a nongame wildlife branch was established in 1983. In 1986 the department approved a policy statement on urban wildlife management by the AGFD to promote the development and preservation of urban habitat. A comprehensive study of critical and sensitive biological habitats in the Tucson metropolitan area was undertaken, and the AGFD worked closely with the Pima County Planning and Zoning Department to establish procedures requiring developers to prepare a site analysis report with maps of sensitive wildlife habitats and vegetative communities. To help educate the public about the values of integrating wildlife into the planning process, a booklet and a regional habitat map were prepared for Tucson residents indicating several classes of critical habitats identified for preservation. In some special cases, outright purchase of land has been the best option for preservation; in others, developers have been encouraged to design around the most sensitive natural features of a particular site by protecting riparian vegetation, establishing continuous corridors, and disturbing as little natural vegetation as possible. Various tax and density incentives have also been employed to protect critical habitat areas. One of the overarching objectives of Tucson's planners is to develop a corridor of interconnected open spaces for wildlife and people based upon the riparian habitat existing throughout the Tucson metropolitan area.

URBAN WETLANDS

When discussing the image and values associated with urban wetlands, there have been few studies to determine people's attitudes toward wetlands. One study in Columbia, Maryland (Adams, 1984), posed this question: "Do you think wetlands add to the beauty, diversity, and quality of the human living environment?" An astounding 94 percent of the 600 homeowners surveyed answered positively. There are undoubtedly regional differences in the way in which people view wetlands and wildlife generally, but it is safe to assume that at the very least, attitudes are shifting toward recognition of the value of both wetlands and wildlife on a national, if not an international, level.

The wetlands wildlife habitat closest to the largest population base in the United States is undoubtedly the Jamaica Bay portion of the Gateway National Recreation Area in New York, right at the end of an airport runway, which is also the only wildlife refuge with a subway running through it! This refuge, consisting of 9,155 acres of saltwater bays and marshes, tidal flats, islands, and some upland prairie located between the boroughs of Brooklyn and Queens, provides an invaluable experience for the over 25,000 schoolchildren visiting the Ecology Village, many of whom also camp overnight in an area called the North 40. Over 100 species of migratory birds utilize the refuge during part of the year, and this urban wetland is an important breeding ground. Sixty percent of the fresh water entering the Jamaica Bay wetlands is effluent from sewage treatment plants, and the remainder is stormwater runoff from city streets and the runways of Kennedy Airport. Over 200,000 people visit the refuge annually.

During a survey of natural resources by planners in Portland, Oregon, it was discovered that more than 4,000 wetland acres existed within the city, with an unusual abundance of wildlife ranging from coyotes and bobcats to bald eagles and peregrine falcons. Even elk have been spotted in the wetlands, leading the city to establish an official urban wildlife refuge at Oaks Bottom that also is a principal feeding territory for a colony of great blue herons. Appropriately, Portland adopted the marsh-loving heron as the official city symbol in 1987. Other urban wetlands in the Portland/Vancouver, Washington, metropolitan area are also being preserved and integrated into a regional wildlife system linking a mosaic of natural areas and administered by a Metropolitan Service District supported by four counties and many municipalities.

Two other notable major wetland wildlife refuges in metropolitan areas are the Tinicum Marsh on the outskirts of Philadelphia, which is operated by the U.S. Fish and Wildlife Service, and Bayou Sauvage within the city of New Orleans, Louisiana. Bayou Sauvage encompasses more than 18,000 acres of wetlands and is a wintering ground for 90,000 ducks and coots; bird such as glossy ibis, snowy egrets, and great blue herons nest in the wetlands.

Constructed wetlands, both for stormwater and for wastewater treatment, can be integrated into the open space of new housing developments and other areas in such a manner as to provide wildlife corridors and reserves but also valuable open space. The NIUW initiated research in 1982 to determine the extent of wildlife use of different types of stormwater control basins; the study found that the typical grass detention basins were little used by wildlife. Permanent water impoundments with a variety of aquatic plants, on the other hand, provided habitat for waterfowl and other wildlife. It was also determined that shallow ponds with gently sloping sides, less open water, and more aquatic vegetation provided more food and cover for wildlife than did deep, steep-sided ponds. One of the more interesting discoveries was that soil deposited from upstream erosion typically formed *sedimentation bars* above the water surface at stream inlets in shallow ponds; these bars, along with the shallow surrounding water, provided attractive feeding and resting areas for waterfowl, marsh birds, shorebirds, and other wildlife. The study recorded six species of mosquitoes in the ponds, but also found that all ponds had large numbers of natural control agents that kept mosquito populations in check. Among the predaceous aquatic insects recorded were mayflies, dragonflies, damselflies, water scorpions, diving beetles, backswimmers, water striders, giant water bugs, and water boatmen. Most impoundments also contained populations of sunfish, mosquito fish, and fathead minnows that feed on mosquito larvae and pupae. Based upon this study and others, the Institute prepared a booklet entitled *Urban Wetlands for Stormwater Control and Wildlife Enhancement* (Adams and Dove, 1984a) which recognized the benefits of designing stormwater control basins to provide storage, treatment, and wildlife habitat.

A number of constructed wetlands have been designed for either stormwater or wastewater treatment, with the additional objectives of providing wildlife habitat and public bird-watching opportunities. As previously mentioned, virtually all wetlands will attract wildlife, regardless of whether they are specifically designed for that purpose, but with additional consideration related to the design and selection of plant species, the attractiveness of the wetlands as a habitat can be greatly increased. Among the facilities designed from the very beginning to attract wildlife are the Mt. View Sanitary District marsh system in Martinez, California; the Columbia, Missouri, Eagle Bluffs Wildlife Area; the Orlando, Florida, Easterly Wetlands; the Arcata, California, system; the Sacramento County, California, demonstration wetlands; and the Show Low–Pinetop, Arizona, lake and wetlands facility. Many others, such as the constructed wetland wastewater polishing system completed in 1993 in Beaumont, Texas, were not deliberately designed with the objectives of providing wildlife habitat and interpretive elements, but they have naturally evolved in that direction due to the attractiveness of the wetlands to bird species and the associated interest in such areas by bird watchers. The Beaumont "Cattail Marsh" attracted the interest of environmentalists and wildlife professionals across Texas and resulted in the development of a variety of research and recreational activities related to the wetlands. A detailed description of the Beaumont facility, as well as some of the others mentioned, is presented in Chapter 10.

THE IMPORTANCE OF INTERPRETIVE SIGNAGE

Interpretive signage, educational brochures, and trail guides have been developed for many of these facilities, ranging from poorly designed and relatively primitive to fairly sophisticated in both design and graphics. Some examples are shown from the Orlando Easterly Wetlands (Figure 8.3). An excellent example of interpretive

Figure 8.3 *Orlando Easterly Wetlands interpretive signage (Robert Knight).*

Figure 8.4 *Boyce Thompson Arboretum, interpretive signage (Craig Campbell).*

signage (Figure 8-4). is from the marsh and lake at the Boyce Thompson Arboretum in Globe, Arizona. Another (Figure 8-5) is from the Woodland Park Zoo in Seattle.

The role of wetlands in purification of water is conveyed through a sign posted at a pilot scale constructed wetland treatment system at the south side of the Fort Collins, Colorado, wastewater treatment facility (Figure 8.6).. The positive contribution that well-designed graphic signage can provide to a wetlands project is usually underestimated and underfunded. Although good interpretive signage costs

Figure 8.5 *Interpretive signage, marsh at Woodland Park Zoo, Seattle (Craig Campbell).*

Figure 8.6 *Fort Collins wastewater treatment plant. Interpretive signage at pilot-scale constructed wetland cells (Craig Campbell).*

more than the typical wood signs with lettering routed out in a maintenance shop, the combination of a well-crafted "story line" with professional graphics conveys a message with more impact and meaning. There is considerable room for improvement in the quality of most interpretive signage and graphics, and clearly a need for more involvement by professional graphics and interpretive design specialists in the development of signage.

Both lakes and ponds, along with constructed wetlands, can be designed and managed to either attract or inhibit visitation and use by particular wildlife. It must be noted that attracting waterfowl in large numbers to a pond or lake can cause significant overloading with nutrients that may offset to some degree the renovation objectives of wetlands and ponds and that may affect the overall water quality.

It cannot be disregarded that within certain urban settings, some forms of wildlife are legitimately viewed as pests by many people. Thus, careful consideration must be given to the impact of building habitats that may attract some of these nuisance species.

The education of city and suburban residents on issues of wildlife and ecology is essential; so, too, is the proper preparation of park and wildlife managers to understand ecological concepts and to incorporate them into a multiple-use scheme compatible with other urban activities. Care must be exercised in controlling nuisance or hazardous wildlife in ways that meet with public acceptance.

One would hope to see the development of increased public understanding of the ecosystems we inhabit, along with policies to nourish and support the protection and development of a healthy and balanced wildlife population. It is well established that riparian zones generally feature the most diverse range of species to be found in any landscape; in arid climates, the contrast between riparian and surrounding zones is even more dramatic. The zones between two different types of ecosystems, called *ecotones* usually support the widest range of species. In the constructed landscape, we are creating artificial ecotones that connect our constructed wetlands with the surrounding landscape. Therefore, it is crucial that these connective elements be carefully considered if wildlife and plant diversity are objectives. Edges and corridors are typical patterns for these connective elements; they may be only several feet wide but extremely long. In any event, the design and planning for human connection to the natural landscape and to habitat areas is critical. In some cases, trails and other public access elements may need to be far removed from critical habitat areas that attract wildlife species sensitive to human intrusion. Most species within urban settings, however, have adapted to humans in close proximity, providing ample opportunities for visitors to view blue herons, ducks, redwing blackbirds, and amphibians.

Children, in particular, are intrigued with pond ecology and the concept of plants assisting in the breakdown of their own wastes. The Los Padillas Elementary School outdoor classroom and constructed wetland project in the south valley of Albuquerque has been a widely publicized success story that many other schools are now attempting to replicate. (See Chapter 10 for a full description of this project.) Public education on environmental issues has produced a much higher awareness of the value of both wetlands and wildlife over the past decade or so, and most surveys of public opinion now indicate a much more favorable attitude toward these elements than they did in the past.

DESIGNING FOR WILDLIFE

While surface water ponds are not always included in constructed wetland projects, they are essential to provide the aquatic habitat required to attract a wide variety of wildlife. Principles of waterfowl habitat design have been developed over the years. They are used by the U.S. Fish and Wildlife Services and other agencies involved in creating waterfowl nesting sites. These efforts involve the creation of islands, bays, and particular aquatic plantings that provide maximum food and cover. In addition, artificial rafts and "loafing" platforms can be provided if the pond or ponds are of sufficient size.

The NIUW provides the following design guidelines for stormwater impoundments to favor wildlife habitat:

Where possible, impoundments for stormwater control should aim to retain water rather than merely detain it.

Pond design must meet applicable stormwater control criteria, including legal requirements.

Natural resources personnel, including biologists, should be consulted during the planning and design stages.

All potential pond locations should be evaluated to select the most suitable site in relation to the developed area and surroundings, and in recognition of physical, social, economic, and biologic factors.

There should be an adequate drainage area to provide a dependable source of water for the intended year-round use of the pond, considering seepage and evaporation losses.

The soil on site must have sufficient bearing strength to support the dam without excessive consolidation and be impermeable enough to hold water.

The pond site should be located in an area where disturbances to valuable existing wildlife habitat by construction activities will be avoided or minimized.

For maximum wetland wildlife value, water depth should be from 15 to 24 inches for 25 to 50% of the water surface area with about 50 to 75% having a depth of approximately 3 1/2 to 4 feet. A greater depth may be advisable for more northern areas subject to greater ice depths. A side slope of 10:1 or less is preferable to steep slopes for wildlife use. Shallow ponds in our study (average water depth 2.3 feet with average side slopes of 16:1) were superior to deep ponds (average water depth 6.8 feet with average side slopes of 3:1) with respect to wildlife use. Also they are safer for children who might wade or fall into the ponds.

Ponds should be designed with the capability to regulate water levels, including complete pond drainage, and with facilities for easy cleaning, if necessary.

For larger ponds (those approximately 5 acres or greater), one or more small islands are recommended. The tops of the islands should be graded to provide good drainage. Appropriate vegetative cover should be established to prevent erosion and provide bird nesting cover. (Adams and Dove, 1989)

The most critical component, however, of a well-managed wetland for wildlife is water level manipulation from season to season. This level of management is often not achievable given limited personnel and lack of means of control within the aquatic system. Where there are resources for intense management of a wetland, there should be a program for regular drawdowns that will encourage diversity of vegetation. A thorough treatment of the subject is offered by Neil F. Payne, Professor of Wildlife at the University of Wisconsin, in *Techniques for Wildlife Habitat Management of Wetlands* (Payne, 1992). This level of management is not option in most cases, and water level and vegetation management may be left to chance for many years with no intervention. Such lack of control may not significantly affect wildlife use in many areas, particularly those with fewer aggressive plant species that are considered pests in other locales.

Aquatic plants form a very high percentage of the diet of many species of waterfowl. Among the most useful plants as waterfowl food are pondweeds, wild rice, and wild celery. In some cases, such as duckweed, the entire plant is eaten; in others, only the seeds or tubers are utilized.

Constructed wetland treatment systems provide more species richness and higher population densities of birds than other wetlands (McAllister, 1992, 1993a, 1993b). The total number of bird species observed at six different constructed wetland sites located in the arid West, Florida, and Mississippi ranged from thirty-three to sixty-three. It is interesting to note that this inventory documented the highest total bird densities at arid sites in Arizona and Nevada where there are fewer alternative natural wetland habitats.

In order to increase the attractiveness of a constructed wetland to birds, nesting and roosting boxes along with feeding stations can be installed. In ponds that are large enough to harbor waterfowl, islands or floating nesting platforms tends to attract more birds than areas not protected from predators. Construction of birdhouses is an ideal volunteer activity for Scouts, school classes, and community organizations. The local chapter of the National Audubon Society can provide detailed information on the proper dimensions of nest boxes and entrance holes for a variety of birds. Shelters for squirrels are also quite easy to construct and can be made from nail kegs fitted with a top and an entrance hole; folded tires have been used successfully for the same purpose. An annual maintenance program is needed to remove unwanted debris and to ensure that the structures will be regularly utilized.

Rock or log piles can be placed so as to provide cover for chipmunks and other wildlife. They can be arranged to be both attractive and functional.

Many schools, particularly elementary schools, are attempting to utilize existing natural areas or unused portions of their property for environmental studies by developing ponds, native plantings, trails, and outdoor classrooms. One such example, initiated by a local landscape architect and the school principal, is the Washington Environmental Yard, which was largely created with volunteer community labor. An existing 1.5-acre asphalt play area was redeveloped into an outdoor classroom and wildlife habitat by removing all asphalt and developing a half-acre natural resource area consisting of ponds, wooded areas, and meadows for use in studying and teaching environmental and science classes.

Another excellent example of the transformation of leftover school property into wildlife habitats and outdoor classrooms for environmental education is at the Crestview Elementary School in Boulder, Colorado. Under the leadership of Debbie Kammerer, an outstanding native plant and wetland habitat was created along a small channel naturally fed by springs and seeps. Among the features of the project are an outdoor classroom, a wetland boardwalk, trails, and a wide diversity of both wetland and upland native plantings (Figure 8.7, 8.8, 8.9).

The authors were responsible for developing a similar project for Los Padillas Elementary School in the South Valley near Albuquerque (Figure 8.10 and 8.11). The Los Padillas Wildlife Sanctuary, as it is now called, is a 4-acre site adjacent to the school that had been preserved in a natural state, with native cottonwoods and grasses along with introduced trees such as Russian olive and tamarisk. The

Figure 8.7 *Crestview Elementary School—signage at the entry to the wetland habitat study area sign (Craig Campbell).*

Figure 8.8 *Crestview Elementary School wetland habitat boardwalk* (Craig Campbell).

site was developed with a constructed wetland for treating the school's wastewater (replacing a sand mound system), a naturalistic pond receiving UV-disinfected effluent from the wetland, an outdoor classroom seating up to fifty students, an interpretive trail, four different native plant community exhibits, and stone benches.

An organization mentioned previously in this chapter, the NIUW, is a scientific and educational organization that advocates the enhancement of urban wildlife values and habitat. As the only national conservation organization with programs

Figure 8.9 *Crestview Elementary School wetland outdoor classroom* (Craig Campbell).

Figure 8.10 Los Padillas Wildlife Sanctuary—entry gate and sign (Craig Campbell).

dealing almost exclusively with fish and wildlife in urban, suburban, and developing areas, the NIUW is an excellent source of technical services, publications, research, and education. One of their publications mentioned previously in this chapter (Adams and Dove, 1984) addresses the means of designing stormwater basins for optimal wildlife habitat; others consider the pest problems sometimes associated with certain wildlife species.

Figure 8.11 Los Padillas Wildlife Sanctuary—entry gate and sign (Craig Campbell).

9

Art, Engineering, and the Landscape

Only through art can we get outside of ourselves
and know another's view of the universe which is not the same as ours
and see landscapes which would otherwise have remained unknown
to us like the landscapes of the moon.
Thanks to art, instead of seeing a single world, our own,
we see it multiply until we have before us
as many worlds as there are original artists.
—*Marcel Proust* (*The Maxims of Marcel Proust,* 1948)

The reader may wonder what a chapter on art is doing in a volume on constructed wetlands. The answer is that art, in the sense of an aesthetic approach, and artful design should be fundamental values in every element built in the environment. In addition, we strongly feel that artists have a major role to play in interpreting both constructed and natural features of the landscape and in helping to develop deeper, even subconscious connections with, and understanding of, the biological processes occurring in wetlands, as well as other features of constructed and natural environments. That this is, in fact, occurring in various projects around the country may not be readily apparent at this time; but as this chapter will illustrate, there are some exceptional examples of artists working in collaboration with landscape architects, biologists, engineers, and others that deserve wide recognition.

One of the more unfortunate developments of the industrial age has been the divorce of engineering from architecture and art. It is abundantly clear from history that early engineering innovations or inventions often originated with artists such as Da Vinci, and that "artful" influences played a major role in engineering into the early twentieth century. Examples abound: the Eiffel Tower, the Brooklyn Bridge, and the Golden Gate Bridge.

Industrial design, if one can utilize that term as a catchall for many elements that constitute the urban fabric, conveyed a keen sense of place and a creative use of materials up to about the 1940s in the design of street lights and poles, park shelters, bridges, fences, gates, and many other elements now devoid of either art or contextual design. Unfortunately, the period from the 1960s to the 1980s was one of minimalist abstract public art with which most of the general public felt no connection. The changes that have since occurred within the field of public art projects, as represented by some of the most progressive programs in the country in Seattle, Portland, Minneapolis, and Phoenix, have rejuvenated the public's connection with these artists, with their creative imagination and site-specific, locale-related artworks. Many wonderful public art projects developed nationwide over the past ten years or so have reestablished the crucial sense of connection between art and the viewer, with a panoply of expressions in sculpture, murals, plazas, and other artworks often well integrated with the work of landscape architects.

ART AND THE ENVIRONMENT

It can easily be argued that historically, most civilized cultures never developed the concept of art as a practice, profession, or value separate from everyday living. Indeed a majority of the normal articles of daily use in earlier cultures—pottery, tools, clothing, adornments—received a level of attention and artistic care in their fabrication that far exceeded their functional requirements. Even in the written languages of Arabic and Chinese, calligraphy has historically been valued in those cultures as tools capable of expressing important aesthetic qualities. The continued fascination of American architects with the pre-industrial urban and village architecture of Europe, most of which pre-dates the existence of their own profession, illustrates the magnetic and spiritual qualities of environments that themselves are works of art. This attitude, expressed by Frank Lloyd Wright, probably is shared by most architects and landscape architects as well as artists:

> Of this joy in living, there is greater proof in Italy than elsewhere. Buildings, pictures, and sculpture seem to be born, like the flowers by the roadside, to sing themselves into being. Approached in the spirit of their conception, they inspire us with the very music of life. No really Italian building seems ill at ease in Italy. All are happily content with what ornament and colour they carry, as naturally as the rocks and trees and garden slopes which are one with them. Wherever the cypress rises, like the touch of a magician's wand, it resolves all into a composition harmonious and complete. The secret of this ineffable charm would be sought in vain in the rarefied air of scholasticism of pedantic fine art. It lies close to the earth. Like a handful of the moist, sweet earth itself, it is so simple that, to modern minds, trained in intellectual gymnastics, it would seem unrelated to great purposes. It is so close that almost universally it is overlooked by the pedant. (Wright, 1910, p. 21)

A book entitled *The City as a Work of Art* (Olsen, 1986) argues that many of the great cities around the world are deliberate works of art. If that is the case, and I believe it is, then one could argue that people growing up in such environments were enriched *subconsciously* by their visual surroundings, which provided them with an education that we now have closeted into the sterile setting of a classroom. The relegation of art to the museum and gallery, disconnected from our everyday environment, diminishes the more profound experiences that accompany the continuous contact and association with art in the everyday environment.

Asking where we may find a criterion of art, Herbert Read, in *The Grass Roots of Art* (1961), states:

> The answer, of course, is nature. There, absolute and universal, is a touchstone for all human artifacts. And we must understand by nature not any vague pantheistic spirit, but the measurements and physical behaviour of matter in any process of growth or transformation. The seed that becomes a flowering plant, the metal that crystallizes as it cools and contracts, all such processes exhibit laws, which are modes of behaviour. There is no growth which is not accompanied by its characteristic form, and I think we are so constituted—are so much in sympathy with natural processes—that we always find such forms beautiful. The artist in particular, I would say, is a man who is gifted with the most direct perception of natural form. It is not necessarily a conscious perception: he may unconsciously reveal his perceptions in his works of art. Artists are to a considerable degree automata—that is to say, they unwittingly transmit in their works a sense of scale, proportion, symmetry, balance and other abstract qualities which they have acquired through their purely visual and therefore physical response to their natural environment. (p. 19)

Read makes a convincing argument related to the social aspects of art in an industrial age and our modern education system. He writes:

> We have never dared to trace the connections between the disordered state of our civilization and our traditional systems of education. If our schools were producing naturally and normally personalities which we could describe as balanced, integrated or harmonious, we should not be able to tolerate a condition of universal disunity and mutual distrust. We should therefore reexamine our whole tradition of education since the Renaissance and dare to ask ourselves whether it has been generally productive of individual serenity and social harmony. We might then have to confess that in our exclusive preoccupation with knowledge and science, we had omitted to educate those human faculties which are connected with the emotional and integrative aspects of human life—that we had carefully nurtured inhuman monsters, with certain organs of the intelligence gigantically enlarged, others completely atrophied. (p. 111)

These considerations raise much larger questions about modern culture in general than one would normally associate with the design professions and their responsibilities. However, there are opportunities for every one of us, regardless of whether we are involved in academic institutions, consulting firms, or government, to try to influence the process of development toward integration and respect for other visions which can help provide connections between humans and their environment otherwise left unrealized.

NATURE, ART, AND CONSTRUCTED WETLANDS

What is the connection between art and constructed wetlands? The art underlying nature, as exhibited by centuries of botanical illustration, paintings, and photographs, has universal appeal derived from the subconscious recognition of the human connection with both natural forms and natural processes. Writers who have explored the nature of "sacred geometry" and the proportional relationships that underlie the dimensions and forms of some of the greatest architecture of the past also discovered the same forms and proportional relationships in plants. Mathematicians (Kappraff, 1991), architects (Grillo, 1960; Doczi, 1981), naturalists (Thompson, 1917), and others have explored the intriguing mathematical patterns that seem to occur throughout the natural world. Both artists and architects have increasingly been drawn to explore these relationships in their work. In a remarkable series of close-up photographs of plant parts by the German photographer Karl Blossfeldt (1865–1932), some of the basic structures in nature are revealed; Blossfeldt felt that "nature, which constructs and develops untiringly, is not only art, but also, in the sphere of techical skill, is the best possible teacher." There are examples throughout nature of particular geometic patterns, angles, and forms that have been analyzed and found to exhibit, for example, the Fibonacci mathematical series of clockwise and counterclockwise spirals found in sunflowers, certain cacti, and some other plants. (Figure 9.1). The golden mean proportion, as well, is found over and over again in plant forms, and even in fish and other aquatic organisms. In a beautiful example of an artist inspired by such underlying patterns in nature, artists Robert Stout and Stephanie Jurs created a ceramic tile mosaic replication of a Fibonacci series in a walkway and plaza adjacent to the Albuquerque Art Museum (Figure 9.2).

FLOWFORMS AND WATER QUALITY

Possibly one of the first efforts to combine art, water treatment, and biology in a singular and unique sculptural concept was that of a hydroengineer named Theodor Schwenk. Schwenk founded the Institute für Strömungswissenschaften (Institute for Motion Research) in Germany over fifty years ago and undertook research with a mathemetician, George Adams, which led them to correlate the curved vortex or

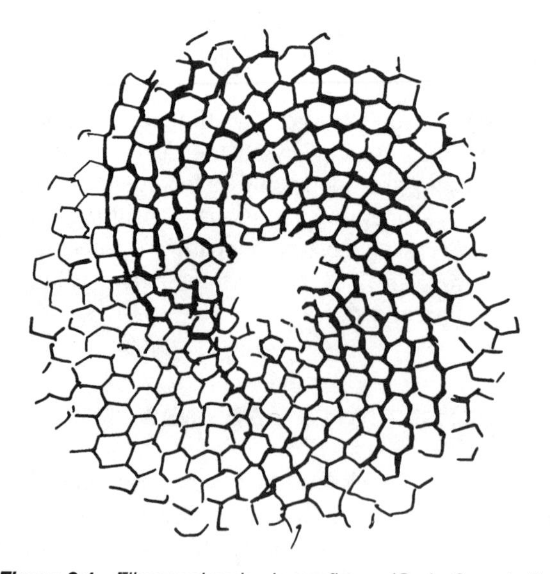

Figure 9.1 Fibonacci series in sunflower (Craig Campbell).

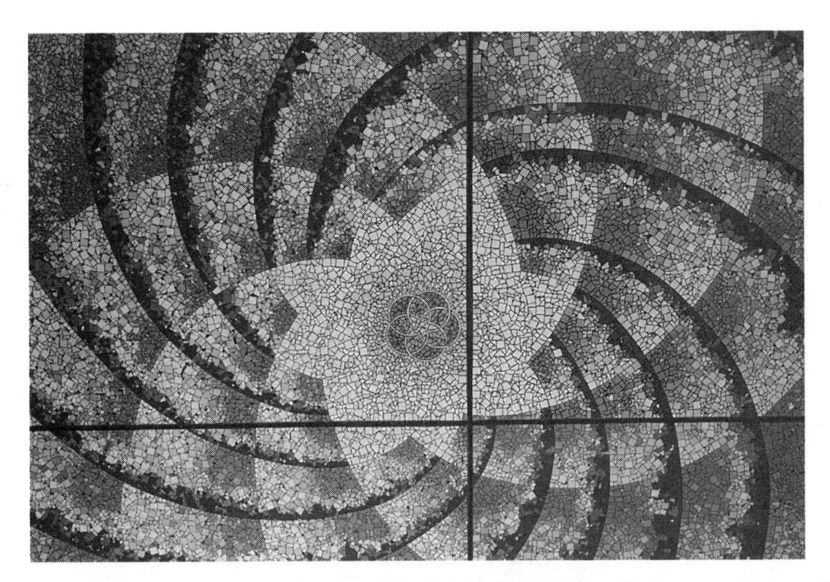

Figure 9.2 *Fibonacci series in mosaic artwork by Robert Stout at the Albuquerque Museum* (*Robert Stout*).

funnel shape prevalent in movements of water and in the morphology of plants. In a book entitled *Sensitive Chaos* (Schwenk, 1965), the author addresses the "creation of flowing forms in water and air" by illustrating, in detail, the similarities in the patterns that develop in the movement of both water and air, particularly spirals and vortices, and the universal depiction of these forms in art (Figure 9.3). An English sculptor, John Wilkes, used the concepts of Schwenk and Adams in developing a series of vertebra-shaped symmetrical forms on an inclined plane, designed to channel water with a pulsating meander. The result was a figure eight movement of water within a series of offset bowls, with water circulating clockwise on one side and counterclockwise on the other—a movement Wilkes later termed a "vertical meander."

In 1973, the *Flowform* design, as it was later called, was installed as part of the sewage treatment system at the Rudolf Steiner School in Järne, Sweden. The head of the school, Arne Klingborg, wanted to "tackle the question of waste water processing not only in an aesthetic way, but in a functional way so that this would be a beautiful situation which people enjoy . . . which also nature enjoys. It's led to the most beautiful gardens you can imagine which are also a sewage plant (Carde, 1990)". The system, which has been in operation for over twenty years, consists of a primary treatment lagoon, seven ponds, and three Flowform cascades to treat all wastewater from the facility which serves 200 students and faculty (Figure 9.4). The biological life at each stage of the treatment process grows more complex from beginning to end, with anaerobic bacteria in the lagoon; the next ponds in the system provide an environment for aerobic bacteria, various forms of algae, and other microorganisms; and the final pond contains fish and a rich range of plants and animal life. Although the Flowforms do oxygenate the water, advocates of the system insist that the rhythms and patterns of the water, as enhanced

Figure 9.3 *Pattern of water flow in nature (Craig Campbell; adapted from Schwenk, 1965).*

by the Flowform proportion and shape, also create a special environment that supports a richer range of microorganisms providing the cleansing effect. They also state that Flowform water stays fresh longer than tap water.

To my knowledge, no definitive tests have been made to determine whether wastewater receives additional benefits from being channeled through Flowforms; but Prof. Gary Coates of Kansas State University, who has visited the school in Järne, reports that not only is the Flowform odor free, but that the entire wastewater processing system provides a beautiful element of the school's landscape. Jennifer Greene, founder of Waterforms, Inc., in Blue Hill, Maine, has developed Flowform systems in the United States and is continuing to study their potential application. One of her installations, at a lagoon in San Lorenzo Park in Santa Cruz, California, is said to have contributed greatly to eliminating algae problems in a closed pond with heavy duck use.

Figure 9.4 *Flowforms (Tim Hough; from Hough, 1984. Courtesy of Michael Hough, Hough Woodland Naylor Dance Leinster).*

ECOLOGICAL ART

Engineers, architects, landscape architects, and others involved in the design of public facilities are increasingly recognizing the advantages of incorporating the talents of artists, biologists, and others to assist in adding other dimensions and interpretive elements to their projects. In addition, many artists have decided to dedicate their work to revealing processes in nature and to connecting humans with their environment through their art.

There is increasing interest and involvement by artists in ecological art, as is well documented in the book *Fragile Ecologies* (Matilsky, 1992), which was prepared in conjunction with an exhibit of the same name. Mel Chin, in *Revival Field,* utilized six species of plants within a chain link enclosure on a contaminated landfill site in St. Paul, Minnesota, to explore the potential of the plants to remove cadmium and other heavy metals. Herbert Bayer, in his *Mill Creek Canyon Earthworks* project in 1982, created a park with sculptural earth forms that also functioned as stormwater drainage basins in King County, Washington. As an example of the underlying philosophy of some of the artists interesting in creating linkages between people and their natural environment, the sculptor Ned Kahn of San Francisco describes his personal objectives thus:

> I strive to create environments and sculptures that enhance people's awareness of natural phenomena and capture the mysteriousness of the everyday world. . . . My large-scale works were developed in collaboration with various architects, landscape

architects, engineers and scientists. My current work focuses on making invisible forces in the environment visible.

Many of his works incorporate flowing water, fog, wind, sand, and light in complex systems that are intended to change over time. An example of his work is *Basin of Attraction* at the Artpark in Lewiston, New York, a spiral-shaped granite basin that creates a whirlpool in the Niagara River (Figure 9.5).

Again, to bolster my thesis that many artists are becoming increasingly concerned about our degraded environment and need to create works that both help educate and effectuate change, David Middlebrook of Los Gatos, California, expresses his objectives in similar terms:

> A work of art which responds to a specific environment should assist and stimulate information related to this space. My work is an attempt to present the diverse and fragile nature of natural materials, and to illustrate man's delicate and inevitable relationship to his environment and to this planet. As we develop greater knowledge and technology, allowing us to unravel the mysteries of our universe, we must evolve our consciousness and sensitivity to the fragile nature of this delicate balance. It is my view that the long-term stewardship of this planet will be equally the responsibility of our poets and engineers.

Figure 9.5 Basin of Attraction *by Ned Kahn (Ned Kahn).*

This, of course, represents precisely the philosophy and the objectives being promoted by the author, and will ideally become a cultural norm in the future.

One of the most innovative attempts in this century to combine elements of engineering, public education regarding wastewater treatment processes, and art was developed by the Phoenix Arts Commission in 1991. In a commendable effort to focus the attention of people living in the most arid zone in the country upon issues of water conservation and treatment, the Commission developed a unique call for entries described as follows:

WATER ISSUES AND PUBLIC PERCEPTION

The Phoenix Arts Commission's Percent for Art Program is seeking an artist or artist team to engage the public interest in the critical issues of water supply and wastewater treatment through the design of a public tour of Phoenix's 23rd Avenue Wastewater Treatment Plant. The tour's audience will be both adults and schoolchildren, and will include presentation and interpretation of the water treatment process and other pertinent water issues. It should also commemorate the employees who operate and monitor the plant's processes. The goal of this project is to create a tour and education program that encourages dialogue on water supply and recycling issues, and begins to change the public perception of wastewater as unfit for any use.

In a desert city such as Phoenix, Arizona, water supply is a topic of primary concern. So far, the concern is prevalent among city officials and local water suppliers. The public information process has begun, but the mandatory water conservation measures have not yet been instituted. The public perception has yet to catch up with the natural, economic, and political realities that limit the future water supply for a growing population.

At the same time, the population's wastewater is increasing, and new technology, along with new EPA standards, provides an increasingly usable wastewater product. Technically, state-of-the-art treatment of wastewater provides potable water, but once again, the public perception has not kept up with the implications provided by the new technology. Much education needs to occur before the public can fully use the potential that this "new" source of water presents (Phoenix Arts Commission, 1991).

This call for entries in a public art program established a new standard that was light years ahead of most public art programs, which typically were more focused on decorating a particular building or site. The prospect of other municipalities emulating the example of Phoenix in melding the function of the artist with that of the engineer, the operator, and others within an educational and interpretive setting is truly exciting. The potential to combine the talents of the artist, the landscape architect, the biologist, and the engineer in efforts to convey both the beauty and the natural functions of constructed wetlands has not yet been fully explored, but I am confident that we will see the fruits of such efforts in the very near future. As Bob Thayer (1994) pointed out:

Artful interpretation is necessary to offer alternative visions and to explore and make sense out of the unseen. Bringing core ecologies to the surface will be an important role of landscape artists and designers. The continually unfolding complexity of the

> natural world and the inability of traditional forms to represent these changes will result in the evolution of new, unfamiliar surface-core relationships. A critical function of landscape architecture and environmental art will be to continually interpret the relationship of human beings to their environment in spatial, visual terms. Since the change in relationship between people and nature is accelerating, new formal interpretations are required at an ever-increasing rate. Here is where artistic creation plays a key role. Art has the ability to anticipate society. Genuinely artful interpretation offers a range of possible futures by which sustainable landscapes can be identified, emphasized, evaluated, and made visible. (p. 316)

In Phoenix, the result of the call for entries was the selection of an artist, but unfortunately, the project did not proceed due to a complete lack of support by the city department responsible for wastewater—a completely different attitude than that of the Director of the Phoenix Department of Public Works, Ron Jensen. Most truly innovative efforts, particularly within public agencies, originate with individuals who have both imagination and perseverance, often meshing in a symbiotic manner with others who are in a complementary position to effectuate change. In Phoenix, Deborah Whitehurst, director of the Phoenix Arts Commission, along with Gretchen Freeman, then director of the city's public art program was instrumental in moving public art into the realm of urban planning and public education. In 1988, Ms. Whitehurst commissioned a master plan for public art from William Morrish and Catherine Brown, who direct the University of Minnesota's Design Center for American Urban Landscape. The plan produced by this team proposed a shift in focus in the public art program to emphasize the city's infrastructure including city streets, bus stops, canals, and water and wastewater treatment plants. The program that followed resulted in over seventy projects enriching every aspect of the public's experience of public facilities. The Phoenix Arts Commission came to believe that ". . . when artists serve as equal members of a design team, they challenge the imaginations of the bureaucrats, engineers, architects, and landscape architects who, in working for municipal government, too often provide only the most standard design solutions. In challenging imaginations, city building becomes more interesting, pleasing, and intriguing. The meaning of a place, which has been excised from most public infrastructure, could be remembered."

An artist was selected to design hatch covers for water valve and water meter boxes as one of a series of projects funded by the Phoenix Water and Wastewater Department. The intention was to place these hatch covers in areas throughout the city where there is a high level of pedestrian activity; existing hatch covers in those areas would be removed and reused in other areas of the city. This project, while not original—Seattle completed a similar project with downtown manhole covers designed by artists in the 1970s—did serve to present an element of art and communication about water and the environment in unexpected locations. The selected artist—Michael Maglich—did a series of hatch covers including one combining relief figures of fossils, piping, and faucets in a design entitled "Arizona Water Faucet Fossil" (Figures 9.6, 9.7).

Unfortunately, the political climate in Phoenix shifted in 1992, and the city no longer supported public art infrastructure projects at the same level as previously. This situation may change for the better at any time, and certainly the accomplish-

Figure 9.6 *Hatch cover for City of Phoenix titled* Arizona King Copper Double Headed Water Faucet Fossil (*Michael Maglich*).

ments—along with the failures—are evidence of an artistic vitality lacking in some other areas of the country.

Papago Park/City Boundary, Phoenix and Scottsdale, Arizona (Jody Pinto, Artist; Steve Martino, Landscape Architect)

In another project supported by the Phoenix Arts Commission, in this case also supported by the Scottsdale Cultural Council, artist Jody Pinto collaborated with landscape architect Steve Martino to develop a concept for a damaged site bounded by major roadways that provides the entrance to Papago Park, a unique natural landmark in the city. The project was intended to heighten the sense of arrival by creating an entrance to Phoenix from Scottsdale and Tempe. In addition, one of Pinto's goals was to re-create the plant and animal life that originally thrived on the land. The concept developed by Pinto and Martino is intriguing due to its ability to provide symbolic meaning, as well as creating a more conducive habitat for reestablishment of native plant and animal life. The basic form is provided by a large treelike structure formed by rock walls and towers that mark the directions of historic and contemporary settlements. Even the symbolism created by the design

Figure 9.7 *Hatch cover for City of Phoenix* Arizona Double-Headed Water Nozzle Fossil (*Michael Maglich*).

has multiple meanings: the form is aligned with the summer solstice, defines the boundary between the cities of Phoenix and Scottsdale, and presents a form related to the Tree of Life—a cultural symbol with particular meaning to local Native Americans, who viewed life as coming into the world through a channel or stem. The "arms" of the structure are aligned to create seven water-harvesting terraces fed by water captured on the uphill slope and channeled by gravity through the central "trunk." This method of capturing runoff water to supply plants was used by the Hohokam people who inhabited the Salt River valley of the Phoenix region and who also developed sophisticated canal systems to channel water to agricultural fields. Martino and Pinto acted as general contractors on the project, which was probably the most efficient means of achieving the desired results with the available budget. Discovering a source of granite "rejects" too hard for a gravel operator to grind up, the designers obtained all of their rock for a mere $2,500 and spent much of the rest of the construction budget for stone masons to construct the walls and towers. This unique project, while not directly related to wetlands, is a worthy example of site-specific artwork developed in a collaborative effort between artists and landscape architects that enhances the environment while providing visual delight. The Papago Park/City Boundary project won an honor award from the American Society of Landscape Architects in 1992.

Phoenix Solid Waste Management Center, Phoenix, Arizona (Michael Singer and Linnaea Glatt, Artists; Black & Veatch, Architects/Engineers)

Although the 23rd Avenue Wastewater Treatment Plant art project was unfortunately canceled, another major opportunity to involve artists in helping to connect the public to a major infrastructure development came along in the form of the $18 million Phoenix Solid Waste Management Center, a transfer facility initially designed by the architecture/engineering firm Black & Veatch. As Ms. Whitehurst and Ms. Freeman developed ideas for incorporating art into projects falling within the domain of Ron Jensen, Director of the Phoenix Department of Public Works, it became clear that the "percent for art" would be a vehicle for opening up the workings of the Waste Management Center to the public and educating visitors on both the problems and solutions addressed by the facility. After two artists, Michael Singer and Linnaea Glatt, were selected for the project and were asked to collaborate, the project really became innovative. Mr. Jensen, a former president of the American Public Works Association, suggested that the artists run the entire project, setting aside Black & Veatch's original scheme but retaining the firm as project engineers. The artists were told to come up with a concept and to "imagine a place" that, as one writer put it, placed the 1 percent for art in charge of the other 99 percent! As Mr. Jensen saw the problem, most public works projects dealing with waste and water treatment are isolated as completely as possible from the community, thus reinforcing the "not in my backyard" image of these facilities. The result of this unique collaborative effort is a facility that incorporates catwalks and an exterior amphitheater, inviting visitors to view its operations (Figure 9.8).

Figure 9.8 *Amphitheatre at SWMF (Michael Singer; photo by O. Stansbury).*

Windows that might otherwise have been ignored are carefully aligned to exploit dramatic vistas both inside and to the surrounding landscape. The artists developed a second phase of the project that calls for bringing in 500,000 gallons per day of effluent from an adjacent treated wastewater channel through a constructed wetland and pond to demonstrate the cleansing effects of the aquatic plants and to restore a desert riparian habitat for plants and wildlife. In addition, a 300-acre site adjacent to the facility has been earmarked for a large-scale research and demonstration project to further educate the public about recyling and re-using materials.

Waterworks Gardens, King County Metro (Lorna Jordan, Artist; Concept Lead; Jones & Jones, Landscape Architects; Brown & Caldwell, Consulting Engineers)

The Pacific Northwest was probably the first region of the country to move beyond the concept of "art in public places" toward "public art," or the concept of the artist playing a major role in a collaborative effort to develop entire projects and sites. Both the City of Seattle and King County have established innovative programs of public art that have provided artists an opportunity to shape stormwater catchment areas, reclaim disturbed landscapes, and work with landscape architects and engineers in possibly the broadest range of art projects in the world. In addition, the Pacific Northwest has nurtured a movement of landscape architects into the area of ecologically based art and design, a logical evolution for a profession with a grounding in both art and natural sciences.

One of the most successful recent collaborative art projects in the country, designed to convey images and information on natural processes, is the Waterworks Gardens, a public art project located at the East Division Water Reclamation Plant in Renton, Washington, just south of Seattle. The artist Lorna Jordan was selected by the King County Public Arts Program to work with the design team responsible for a major expansion of the 80-acre Water Reclamation Plant by Metro, the government agency responsible for both water treatment and public transportation in Seattle and the surrounding King County metropolitan area.

In a rare but marvelous confluence of public funding objectives, Waterworks Gardens represents a unique and creative integration of three objectives of the expansion of the water reclamation plant:

1. The implementation of an arts program within Metro projects.
2. The development of a stormwater treatment facility.
3. The enhancement of an on-site wetland area.

It was the artist who developed the concept of combining the treatment functions, the artwork, and the wetland enhancement into a single publicly accessible project; and it was the artist who, after three years of concept and budget development, was selected to lead the design team for the project (Figure 9.9).

Among the significant features of the project, all of which are viewed by the artist as "physical/conceptual layers," are wet ponds for stormwater treatment; wetland enhancement; public pathways accessible twenty-four hours a day; garden rooms and viewpoints; landscape elements selected and organized by art; and an

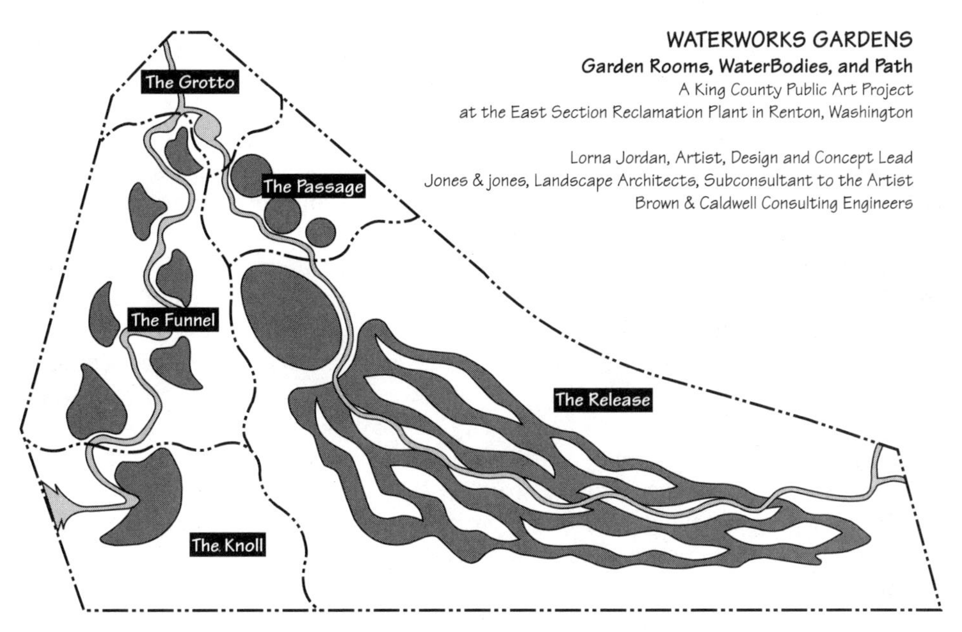

Figure 9.9 *Plan of Waterworks Garden (Lorna Jordan).*

education/research program in which schoolchildren can learn about wetlands, stormwater treatment ponds, and environmental art.

It is somewhat rare for an artist to lead a project of this magnitude, and the scope and success of the final installation are a tribute to Lorna Jordan's effort to significantly broaden the initial program for the stormwater treatment system to include aesthetics, wetland enhancement, public access, and education, as well as engineering. In her concept report, the artist begins her introduction with an appropriate quote from Christopher Alexander's *A Pattern Language* (*1971*):

> As marvelous as the high technology of water treatment and distribution has become, it does not satisfy the emotional need to make contact with the local reservoirs, and to understand the cycle of water: its limits and its mystery.

Lorna Jordan has summarized her project as follows:

> The resulting art project integrates nature, art, and technology while inviting people to observe natural processes of water purification. Waterworks Garden treats stormwater, enhances two on-site wetlands, provides five garden rooms, and creates eight acres of publicly accessible space. The project uses the conceptual framework of "the garden" as the philosophical balancing point between human control and wild nature. Thus, the gardens complement the controlled treatment plant processes while providing a natural, organic way to reclaim water that shifts the balance toward nature. Landforms, plantings, water bodies, and garden rooms combine to form an abstract flowering plant, symbolic of the filtering power of plants to cleanse water. A progression of five garden rooms follows the cycle of the water's purification: impure, working, mysterious, beautiful, and life-sustaining. ("Waterworks Gardens, a Public Art Project, 1990).

Waterworks Gardens is a truly unique functional art project, seamlessly connected and integrated with the surrounding landscape, not the more typical disconnect between the art "object" and its setting. Woven into the project, stormwater is purified as it moves through a naturalistic, open system of terraced ponds, from which it is then utilized as part of the enhancement of the adjoining wetlands. The more intimately scaled garden rooms provide mystery and surprise as visitors follow the water's passage through the site.

The project was funded with $50,000 from the Metro Arts Program and $800,000 from the Plant's Enlargement III construction budget for the wet ponds and wetlands enhancement, for a total of $1.3 million. Completed in 1995, Waterworks Gardens represents the epitome of artists and landscape architects collaborating with engineers to create places for public contact and interaction with some of the basic infrastructure functions of a city normally considered as "out of sight, out of mind" by public officials and engineers.

Along with some of the Phoenix projects, Waterworks Gardens will ultimately be seen as setting a new standard as a project that marks the beginning of a new era in terms of incorporating the visions and unique perspectives of artists into projects otherwise viewed as solely functional and impossible to integrate with interpretive, symbolic, recreational, or educational features. The fact that a wastewater treatment facility expansion project provided $800,000 of the funding for this project is a breakthrough that, one would hope, will establish a precedent for other expansion projects around the country that would normally not consider any public access, art, or interpretation. There needs to be a receptive attitude among public officials, along with a supportive public culture, to achieve the best synthesis of art, landscape, and engineering; and it might be argued that the Northwest as a region provides this milleu better than any other region at this time.

The Water Walk, the main trail through the project, is conceived as a "perceptual and sensory journey which traces stormwater through a series of biological, physical, and visual treatments." The overall design of the trail, the landforms, water bodies, and plantings, are abstractly expressed as a large flowering plant, with visitors weaving through the stems, leaves, berries, and flower, following the path of water purification (Figures 9.10, 9.11). Among the special features along the walk are a series of garden rooms that include The Grotto" (Figure 9.12, 9.13), The Passage, and others. The Knoll provides the initial entry to Waterworks Garden and symbolizes the root of a flowering plant. A formal geometric group of basalt columns is arranged in a forced perspective on each side of the garden room, and water begins the journey at this point by spilling from an iron pipe into a grate-covered channel cutting through the geometry of the columns—a reference to the normally hidden means of dealing with stormwater. The water is released into the open system of ponds at the end of The Knoll's overlook. There is a 40-foot elevation difference between the initial entry at The Knoll and the final garden room, The Release, at the bottom of the hill, where the purified water is released from the final pond to the wetlands and from the wetlands to a creek (Figure 9.14).

Waterworks Gardens establishes one of the most admirable concepts ever developed, within an artistic and symbolic structure, combining actual water renovation with information on water purification and wetland processes. Its development and use will be extremely interesting to all of us engaged in the design

Figure 9.10 *Waterworks Garden, the Release—Renton, Washington. (Lorna Jordan).*

Figure 9.11 *Aerial view, Waterworks Garden, Renton, Washington. (Lorna Jordan).*

Figure 9.12 *The Grotto, Waterworks Garden, Renton, Washington. (Lorna Jordan).*

of wetlands for water treatment who seek imaginative methods of interpreting our projects to the public.

The Pull: Wild Gardens at Paerdegat, Brooklyn, New York (Lorna Jordan, artist; The Portico Group, Landscape Architects)

Another related project that Laura Jordan developed, in collaboration with The Portico Group of Seattle, was a proposal for the New York Public Art Fund's

Figure 9.13 *The Grotto, Waterworks Garden, Renton, Washington. (Lorna Jordan).*

Figure 9.14 *Wetland edge, Waterworks Garden, Renton, Washington. (Lorna Jordan).*

Exhibition *Urban Paradise: Gardens in the City*. Her design, which responded to the rehabilitation of the Paerdegat Pumping Station and the restoration of the Basin wetland areas supported by New York State's Department of Environmental Protection, was to create a series of maintained garden spaces designed to interpret the successional stages that occur as a site is naturally reclaimed. A second parallel set of gardens are conceived as an outdoor learning laboratory where the mosaic of actual successional patterns on the site can be witnessed.

King Street Gardens, Alexandria, Virginia (Buster Simpson & Laura Sindell, Artists; The Portico Group, Landscape Architects/Architects)

The King Street Gardens project is a quintessential example of a collaborative effort involving artists, a landscape architect, and an architect, all equally talented and innovative but each providing a somewhat different perspective and expertise in a truly wonderful meshing of efforts. Becca Hanson, landscape architect of the Portico Group; artists Buster Simpson and Laura Sindell; and architect Mark Spitzer—all from Seattle—were selected in a two-stage invitational competition by the City of Alexandria, Virginia, to design a public park at the center of 75 acres of urban redevelopment surrounding a new Metro station into Washington, D.C. The park site was also intended to provide a symbol as the gateway to Alexandria's downtown historic district. Part of the mandate given the team was to develop a community consensus for a concept that would provide the basis for acceptance and partial funding by the Alexandria City Council. The team, in searching for a symbolic theme that could be woven into the site concept, came up with the colonial tricorner hat as the model for a wire structure topiary to be planted with a living "tapestry" of vines for year-round interest and color (Figure 9.15).

Another major element of the overall park concept, the Sunken Garden, was inspired by small natural cattail marshes bounded by natural rock that the team observed in a nearby area (Figure 9.16). The Sunken Garden is actually a multi-

VIEW OF TOPIARY

CITY OF ALEXANDRIA

KING STREET GARDENS

Buster Simpson ■ Laura Sindell ■ Mark Spitzer ■ Becca Hanson

MAY 1991

2

Figure 9.15 *King Street topiary (The Portico Group).*

purpose constructed wetland designed not only to provide visual interest from the surrounding walks, but also to provide stormwater detention and treatment for the Gardens and for runoff draining from adjacent paved areas. In addition to cattails, yellow iris are to be planted for additional interest around the perimeter of the Sunken Garden, which is formed with brick paving on a slope between the sidewalks above and the wetlands below. This bricked slope is intended to provide an urbane transition between the "people place" above and the "wild place" below and to "reinforce the nature of the enclosed cattail marsh as a bounded element within a cohesive urban framework."

The Hanging Garden portion of the project consists of a trellis structure providing shade to benches located along the base of the trellis. Brick paving, traditional in Alexandria, is used as the material for the surrounding walks on the triangular site, and has been designed to accentuate both the pedestrian circulation system and the drainage patterns of the Gardens (Figures 9.17, 9.18).

Fair Park Lagoon, Dallas, Texas (Patricia Johanson, Artist)

Patricia Johanson has long advocated the involvement of artists in the design and interpretation of basic infrastructure elements of the city, particularly thosed related to water, as well as in landfill and other land reclamation projects; she is dedicated to creating art intended to revitalize natural ecosystems. As Johanson began creating larger works, she became more interested in environmental design, architecture and planning, and became increasingly frustrated at being treated as a "stupid female" when proposing large projects. She therefore decided that she needed an architectural degree to help establish her credibility with the engineers and planners she was determined to work with, and proceeded to put herself through the City

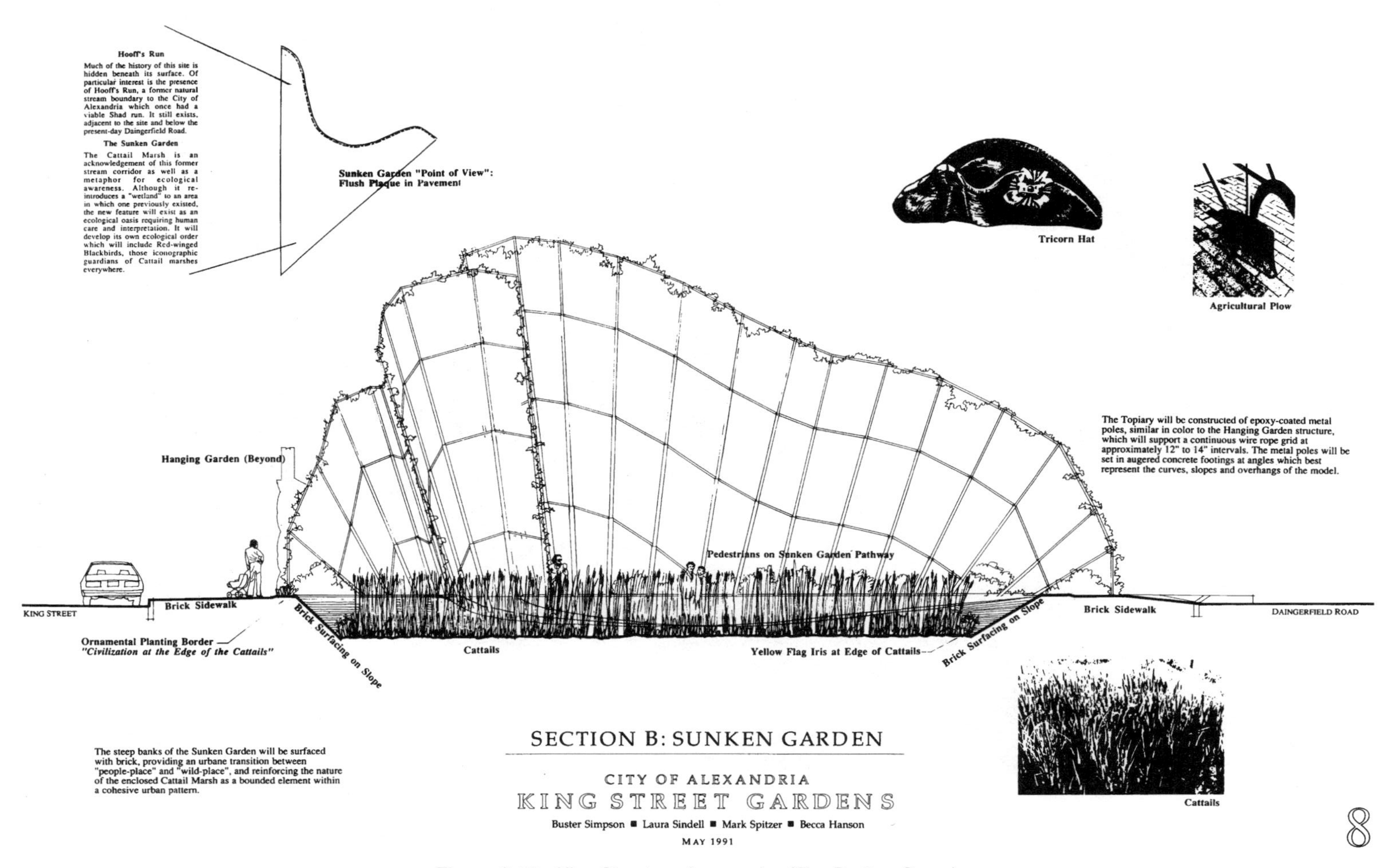

Figure 9.16 *King Street sunken garden (The Portico Group).*

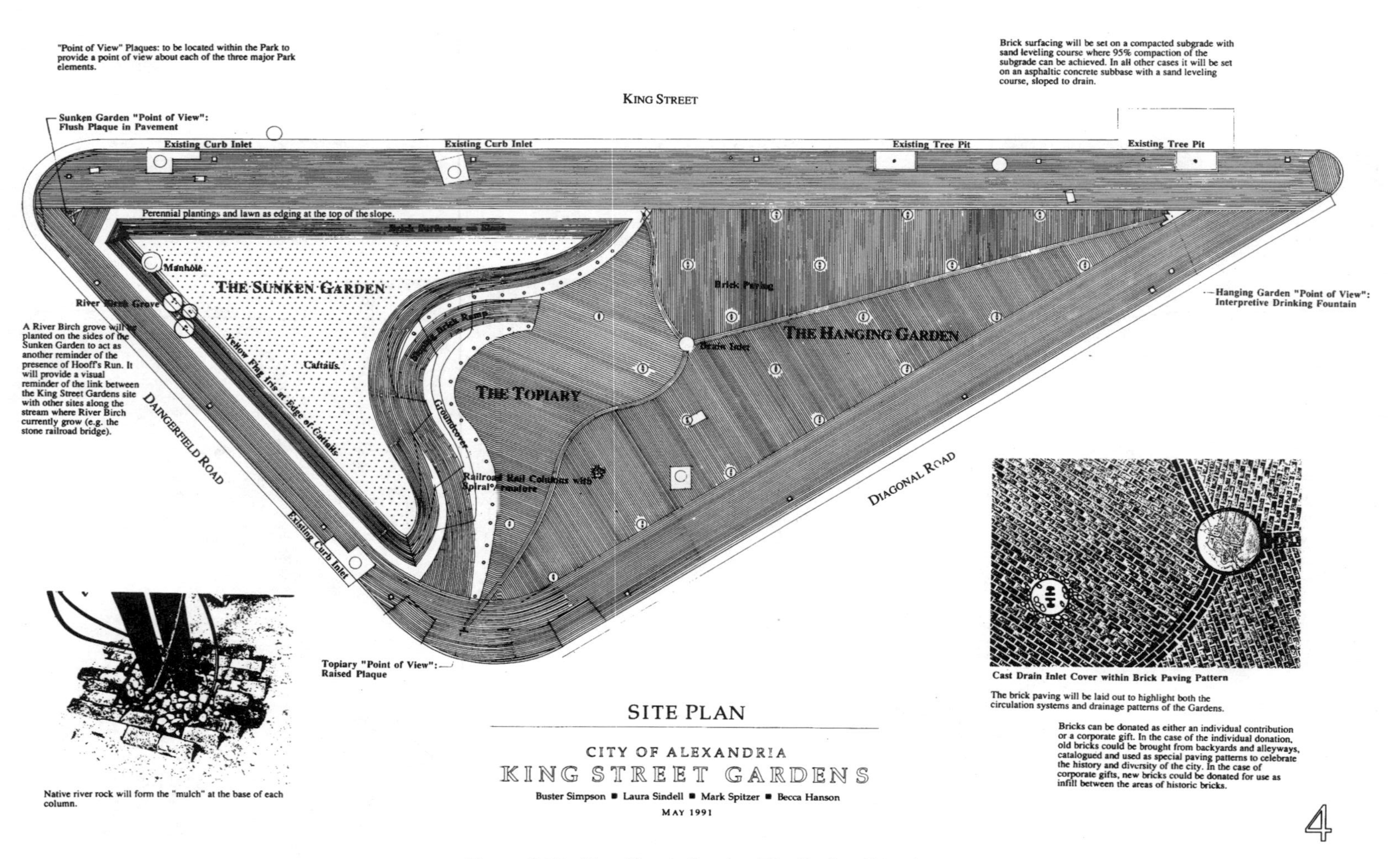

Figure 9.17 *King Street site plan (The Portico Group).*

Figure 9.18 *King Street nearing completion, before cattail planting in the lower basin (The Portico Group).*

College School of Architecture while employed by the architectural firm Mitchell/Giurgola in New York City.

In several innovative projects involving considerable research into wetland processes, Ms. Johanson developed artworks that create both a balanced wetland system and a public educational element. In 1981 Harry Parker, then Director of the Dallas Museum of Art, asked Ms. Johanson to design an "environmental sculpture" for Fair Park Lagoon, a five-block-long lagoon in the middle of Dallas's largest park, and with four major museums along the shore. Faced with the prospect of a lifeless, algae-ridden pond with excess lawn fertilizers washing into the water and exacerbating the problem, Johanson decided to research pond and wetland biology with the objective of combining cleanup of the water with a sculpture that would be an educational tool.

Her sculpture was envisioned from the very beginning as a means of bringing people into contact with plants, animals, and water by creating fanciful paths, bridges, islands, overlooks, and seating; in addition, she deliberately created "housing" for the animals by delineating refuges and microhabitats such as protected islands. The forms that inspired the Fair Park Lagoon sculptures are particular wetland plant leaves and roots. Johanson's concept also incorporated landscaping that served multiple functions, ranging from providing wildlife food to reducing water turbidity and bank erosion. One of the sculpture features, *Saggitaria Platyphylla* (named for the delta duck-potato, a native Texas wetland plant used with this sculpture), provides a mass of interwoven gunited paths in the form of roots as "lines of defense" to break up wave action and to provide access to the water for visitors, placing them within the sculpture and the wetlands (Figure 9.19) Children use the sculptures as a playground for exploration and discovery.

Ms. Johanson feels that "Fair Park Lagoon has become an enormous popular success, and I believe the reason is because it is a many-layered design that responds to 'real' needs—aesthetic, ecological, and functional. It becomes an inclusive, life-supporting, open-ended *framework* that allows for dialogue between art, man, and nature" (Johanson, 1992).

In another sculpture feature, equally multipurpose, Johanson created *Pteris Mulitfida,* a multispan bridge with a form, inspired by a Texas fern, that creates an intriguing landscape of arches and causeways (Figure 9.20). Upon approaching the staff of the Dallas Museum of Natural History with the idea of creating "living exhibits" in the lagoon, rather than inside the museum within glass cases, Johanson received enthusiastic support and proceeded to work closely with the museum staff to develop wildlife habitat plantings on the shoreline and restore the ecosystem of the lagoon. The environmental needs of turtles, fish, birds, and native aquatic plants were researched, and the lagoon was planted with emergent plants such as bulrush, along with wild rice as a waterfowl food source.

Walter Davis, then Assistant Director of the Dallas Museum of Natural History, has commented that "today the Lagoon teems with life. Those who understand the intricacies of a functioning ecosystem find particular satisfaction here. A kingfisher visiting for the first time in decades signals that the water is clear enough for this master fisherman to spot minnows swimming beneath the surface. A pair of lest

Figure 9.19 *Fair Park lagoon,* Sagittaria platyphyll (*Elizabeth Duvert*).

Figure 9.20 *Fair Park lagoon,* Ptereis multifida (*Elizabeth Duvert*).

bitterns, secretive inhabitants of the vegetative shoreline, moved in the first year and has built a nest and raised a family each of the past five years. Ducks and turtles sun themselves on emergent parts of the sculpture, safe from predatory dogs and cats and enthusiastic children. Those plants and animals are not captives held for the enjoyment of human spectators. Most have chosen to live in the Lagoon because it provides food and shelter for themselves and their offspring" (Johanson, 1992, p. 16) This is an admirable testament to the potential for artists to create environments that are valuable both as natural habitats and as a educational and entertaining "people places" (Figures 9.21, 9.22).

As an endnote to an otherwise supportive and enthusiastic description of Patricia Johanson's work in Dallas, I must express some frustration when reflecting upon how difficult it has recently became for any artist to actually create a publically accessible artwork that does not completely comply with ADA requirements, along with all the other applicable local, state, and federal codes related to guardrails, handrails, maximum walk slopes, landings, required widths, surface texture, and every other standard designed to remove any conceivable danger or difficulty to any segment of the population. Unfortunately, such regulations have dampened many innovative ideas that could otherwise provide much more variety, richness, and visual stimulation to our explorations of special places. It is understandable that fairly rigid standards must be applied to the design of public access routes to major public buildings; but logically, there should be exceptions and leeway for special circumstances for artworks that incidentally, also invite public access and involvement. No society on earth has developed as many rigid regulations as ours pertaining to the design of virtually every element in the public—and often pri-

Figure 9.21 *Fair Park lagoon, (Elizabeth Duvert).*

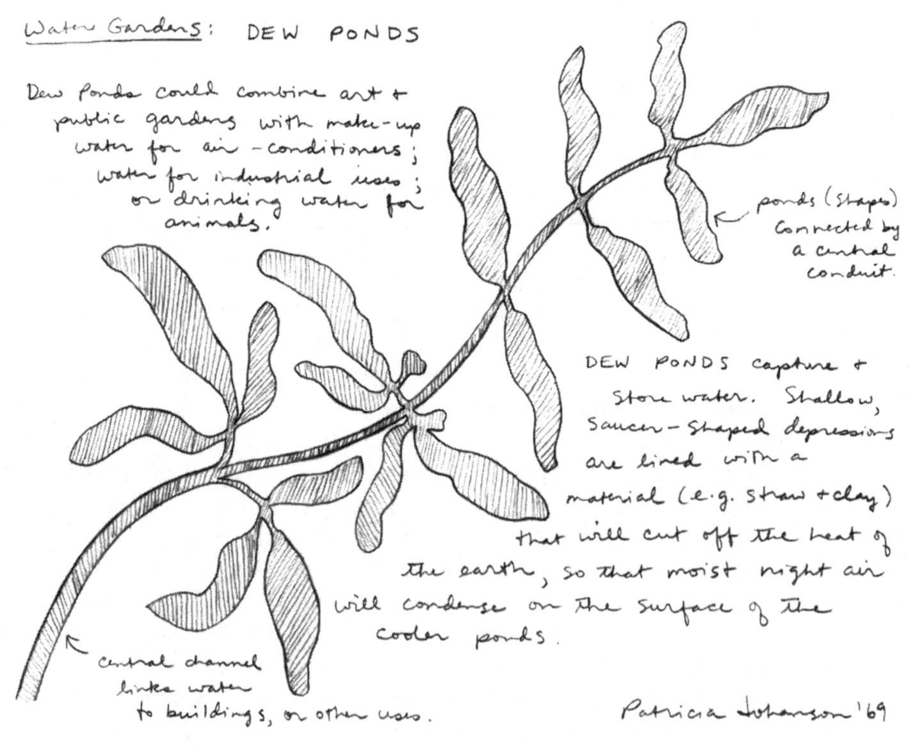

Figure 9.22 *Sketch concept for* Dew Ponds *by Patricia Johanson.*

vate—realm, and the result of such overregulation is a predictable sameness to the appearance of streetscapes, walks, ramps, and other public features all over the country.

WILDLIFE AND ART

It is worth noting that just as engineers and landscape architects are embracing the skills of artists and biologists in some of their projects, some artists are moving closer to a direct involvement in developing works of art connected to wildlife habitat. Other artists have also moved away from monumental ego statements, often insensitive to the environment in which they are created, work that celebrates landscape restoration, biodiversity habitats, and the human connection to specific environments. Again, quoting Bob Thayer (1994), "Indeed, environmental restoration may become the art movement of the twenty-first century, as artists sense a growing social bifurcation between economy and ecology too wide to ignore."(p. 220). The potential for artists to work closely with landscape architects and biologists to convey information on the beauty and the function of constructed wetlands is vast and is only beginning to be explored.

In this country, art has tended to be narrowly circumscribed, defined, and applied in a manner that isolates and divorces the work of artists from processes they should be involved in. This problem is compounded by the rarefied milieu of most art galleries and by the focus of art magazines on gallery-oriented art. While many artists may be somewhat naive regarding the limitations of construction budgets, project schedules, liability exposures, and other issues with which design professionals are familiar, they nevertheless offer the potential of bringing a completely fresh perspective to many projects. In the case of constructed wetland and wildlife habitat projects, we unfortunately do not normally employ the talents of artists in areas where they could bring special life, humor, and beauty to those experiencing the site. There is no reason why artists cannot be more frequently involved in developing concepts for benches, interpretive signage, lighting, amphitheaters, fences, gates, arbors, shelters, and other special features of a project.

Lynne Hull

An artist with a serious commitment to integrating her art into natural ecosystems and creating wildlife habitat is Lynne Hull, an environmental artist of Fort Collins, Colorado, who has created site-specific artworks such as *Stones for the Salmon* for a salmon reclamation project in Northern Ireland. Among her works in Utah and Wyoming are stone carvings, or *hydroglyphs,* which are not only striking compositions, but are also designed to capture rainfall for desert wildlife (Figure 9.23) She has also crafted floating waterfowl nesting islands, bat boxes, pine marten havens, and other wildlife habitat sculptures (Figure 9.24). Ms. Hull consults regularly with wildlife biologists to assist in determining the particular wildlife needs in each area she works in. She was invited by the Wyoming Game and Fish Department to create sculptures along the green belt at Green River, Wyoming, that involved two goose nesting platforms several bat houses, an otter nesting site, butterfly hibernation sculptures, and osprey nesting platforms, all crafted with the

Figure 9.23 Hydroglyph (Lynne Hull).

Figure 9.24 Nesting platform (Lynne Hull).

specific needs of the wildlife in mind and with an aesthetic to appeal to human viewers. By contrast with many of the other artworks mentioned and illustrated in this chapter, much of Ms. Hull's work will be experienced more by wildlife than by humans. Indeed, some of her sculptural creations designed as wildlife habitats or nesting platforms will probably never be viewed by anyone, but this fact does not diminish their value or importance.

Grizedale Forest Sculpture Park

In England, there is a park called Grizedale Forest Sculpture Park, situated within a working timber forest, which also accommodates a variety of wildlife habitat, trails, and campsites. The real attraction of Grizedale Forest, however, is the sixty-five sculptures that have been woven into the forest along a 9-mile trail—the largest grouping of site-specific sculptures in the world. Artists invited to work in the forest on a residency basis are given access to the natural forest's materials and sites. While most artists have been British, several non-British artists have also created artworks in Grizedale, including Lynne Hull. She was given a tarn, or pond, which was in need of both wildlife habitat enhancement and observation stations. As Ms. Hull described her mission,

> One objective of my project was to create a place attractive to humans but also protected by decorative screening to discourage people from disturbing the rest of the area, which would be reserved for wildlife. The installation creates a small place in the habitat where humans can observe birds and other wildlife without interfering in their protected space. With the help of several English artists and other friends, I developed a screened observation point for humans, an island for water fowl nesting, a goose-nesting platform, kestrel and goldeneye nest structures, songbird housing and bat boxes on stilts. The trail into the installation was marked with both real and falsefront birdhouses (Hull, 1993).

Throughout this park, artists have created a wide variety of wonderful elements, each representing a unique perspective and talent, all woven throughout the park as weird surprises, as practical and functional elements, and as wildlife habitat in the form of sculptor-created floating nests for waterfowl. Even the famous master of the ephemeral creation, Andy Goldsworthy, has three major works at Grizedale, one a snake-like clustering of sapling trunks in a work entitled *Sidewinder* that extends for 60 feet on the dark forest floor under spruce and larch trees. Sally Matthews created sculpted but lifelike replications of animals such as a wild boar and a wolf, each one set naturally, as though in motion, into the landscaped setting. The creatures have been formed out of materials from the site, in this case brushwood, heather stalks, and dried mud. Other artists created a wheelchair-accessible path on which visitors encounter sculptural musical instruments powered by wind or hand. Still others symbolically paid homage to ancient ritual structures, traditional stone walls, and even rainfall runoff conveyance devices.

Grizedale Forest is open to visitors for free twenty-four hours a day and is visited by 300,000 people each year. Interestingly, there has not been a single case of vandalism, but instead many examples of visitors helping to repair or extend works.

This is admittedly an unusual example, one virtually unknown in this country, but which represents the incorporation of multiple talents into the enhancement of an otherwise totally natural environment. The encounter with such works of art within this environment has far more impact than the segregated and artificial mounting of displays in galleries and even collections of sculpture carefully arranged on grassy lawns at art shows. One of the only examples of a setting for ecological art, much of it viewed as temporary, is Artpark in Lewiston, New York, on the Niagara River. The artist Ned Kahn created a stone basin near the bottom of a creek feeding into the Niagara and directed the flow through a steel pipe found on the site to create a vortex that drains to the river, illustrated in Figure 9.5. Other artists have also contributed works related to the natural processes of the Artpark site. The objectives of this facility is to encourage aesthetic visions of natural laws that work with the forces of the land. David Katzive, Visual Arts Director in 1975, wrote that Artpark was designed to "change with the seasons, to respond to nature. . . . with this concept in mind we do not expect that artworks or their residue will remain from one season to the next. We expect to relocate, remove, cover over, or allow the natural erosion processes to occur."

Isla de Umunnum, Elkhorn Slough National Estuarine Research Reserve, California (Heather McGill and John Roloff)

There are more artists interested in working directly with environmental interpretation and enhancement than most people might think. At Elkhorn Slough National Estuarine Research Reserve, one of California's last remaining wetlands, Heather McGill and John Roloff created sculptures that enhance habitats for hummingbirds in a work entitled *Isla de Umunnum.* The sculptures are sited on a 5-acre island surrounded by water and a tidal marsh accessible only by a half-mile trail from a visitor's center. In their desire to encourage the survival of native plants and animals, the artists collaborated with the Elkhorn Slough Foundation and sanctuary park rangers in determining the best way to reach their goals. After researching the natural and archaeological history of the site, the artists discovered that hummingbirds were revered by Native Americans and dedicated their project to preserving the birds. Two separate sculptures were created to provide visual interest and to attract hummingbirds. One is the *Trellis,* conceived of as a giant hummingbird feeder, planted with honeysuckle, fuchsia, manzanita, and other native food sources within concentric rings around the trunk of a fallen eucalyptus tree sheathed in copper. The second sculpture, the *Mound,* appears as a half-excavated Native American refuse mound rising out of a semicircular pond to provide the only source of fresh water for wildlife on the island. Heather McGill writes that "The artists engaged in ecological art are our contemporary shamans. Through their work, they attempt to heal the rift that has developed between people and nature. Like many of the early artists before them, they seek to restore balance in a world whose natural vitality is rapidly being sapped" (Matilsky, p. 104).

THE NATURE OF THE COLLABORATIVE EFFORT

Not every project involving artworks will require the collaboration of others, but a significant number of public art projects that require careful integration of the work of the artist and adjacent elements often will. The nature of the collaborative effort between artist, landscape architect, architect, engineer, and public officials is admittedly delicate and at times frustrating. Kay Wagenknecht-Harte, a landscape architect, has provided the best in-depth analysis of the nature of collaborative efforts in her book *Site + Sculpture* (1989), a guide to the collaborative design process. In a certain sense, we are still at the threshold of a new era of collaborative efforts involving more openness, less control by a single "expert," and recognition of the value of new perspectives within the design process. If there is one factor that creates difficulties in collaboration between artists and design professionals, it is the difference in awareness, and thus in attitudes, of professional landscape architects, architects, engineers, and public officials concerning liability, vandalism, and safety issues, along with questions of material longevity. These concerns, which must be taken into consideration by all designers who develop plans for buildings, parks, streets and walks, and other publicly accessible elements, are rarely a major concern of artists, who often may be dismayed at the seemingly paranoid questions raised by designers working with them on a team. In addition, problems often develops when a design team of landscape architects, architects, or engineers is under contract to one municipal agency primarily concerned with "nuts and bolts" and budget issues, while the artist is brought in later to develop a public art project under a separate contract with an arts board. In this situation, two opposing sets of demands and expectations may develop and become difficult to reconcile.

While there have been many highly successful collaborations between strong, creative personalities, there have probably been even more failures derived from ego or personality conflicts; the refusal of an artist to consider issues of safety, liability, and longevity mentioned previously; fickle clients; and problems with public acceptance. A landscape architect understands the potential long-range effects of salts and irrigation spray on certain materials; a structural engineer may have concerns about support systems and wind. As the collaborative process matures, these problems will be reduced, little by little, as artists come to understand the constraints of public art, including issues of long-term maintenance, and as the designers become more open to direct involvement by artists in design issues from the beginning rather than at the end of a project. Of course, the personality issue will always be a major concern. The experience of teams that have worked together successfully in the past should receive consideration over teams put together with little regard for potential ego conflicts. Some have described successful collaborative efforts as not unlike a successful marriage, as the relationship must be sustained over a long period of time and face many frustrating challenges.

As Kay Wagenknecht-Harte (1989) expressed it,

> traditionally, the artist creates an object, the landscape architect an environment, and the architect a building. With the revival of genuine collaborations, the collaborative team has become responsible for creating places—evolving places which animate with the introduction of people going about their daily lives. These special places are *stages* where people can live out special moments, thoughts, dreams, or emotions—all of which are fleeting, but all of which create the sustenance of life. . . . The collaborative projects of today are harbingers for urban environments of the future which will be more responsive to the creative, spiritual, and playful sensitivities of human nature. Heightened sensitivities will be accompanied by increased perceptions, and greater

> demands for designing the built environment more responsibly—an environment which encourages people to think more perceptively, to dream more freely, to feel more deeply, and to laugh more heartily (pp. 123, 124).

As a final footnote to the sentiments expressed by Ms. Wagenknecht-Harte, the need to view art and artists as integral to the understanding, interpretation, and appreciation of our community and our environment is well stated by Lewis Mumford in from *Art and Technics* (1952):

> The work of art springs out of the artist's original experience, becomes a new experience, both for him and the participator, and then further by its independent existence enriches the consciousness of the whole community. In the arts, man builds a shell that outlasts the creature that originally inhabited it, encouraging other men to similar responses and similar acts of creativity; so that, in time, every part of the world bears some imprint of the human personality. Art, so defined, has no quarrels with either science or technics, for they, too, as Shelley long ago recognized, may become a source of human feelings, human values. The opposite of art is insensibility, depersonalization, failure of creativity, empty repetition, vacuous routine, a life that is mute, unexpressed, formless, disorderly, unrealized, meaningless (pp. 139–140).

10

Examples of Multiple-Use Constructed Wetlands

> If man is to remain on earth, he must transform the five-millenia-long urbanizing civilization tradition into a new ecologically sensitive harmony-oriented wild-minded scientific/spirtual culture. . . . Master the archaic and the primitive as models of basic nature-related cultures—as well as the most imaginative extensions of science—and build a community where these two vectors cross.
>
> —Gary Snyder ("Four Changes" in *Environmental Handbook,* G. Debell, ed., New York: Ballantine, 1970)

One of the most fascinating and challenging aspects of the developing art and science of constructed wetlands for water renovation is their ability to be designed in an aesthetically pleasing manner that allows such installations to be woven into the fabric of a community's open space, park, and recreation system. The potential for constructed wetlands to become familiar components of many local landscapes, adding to the overall richness and variety to be experienced on a regular basis, has only begun to be explored. One of the best descriptions of the qualities that make particular landscapes memorable is provided by J. B. Jackson in *The Necessity for Ruins* (1980):

> This is how we should think of landscapes: not merely how they look, how they conform to an esthetic ideal, but how they satisfy elementary needs: the need for sharing some of those sensory experiences in a familiar place: popular songs, popular dishes, a special kind of weather supposedly found nowhere else, a special kind of sport or game, played only here in this spot. These things remind us that we belong—or used to belong—to a specific place: a country, a town, a neighborhood. A landscape should establish bonds between people, the bond of language, of manners, of the same kind of work and leisure, and above all a landscape should contain the kind of spatial organization which fosters such experiences and relationships; spaces for coming together, to celebrate, spaces for solitude, spaces that never change and are always as

memory depicted them. These are some of the characteristics that give a landscape its uniqueness, that give it style. These are what make us recall it with emotion. (pp. 16–17)

Jackson also points out that at present we are groping for new kinds of community, that old familiar landscapes are disappearing, and that new landscapes involve new relationships. The purpose of this volume has been to explore one relatively small element of the overall landscape, with an emphasis on conveying information on the potential for solving at least some wastewater and stormwater problems while at the same time providing other important functions that can be integrated into the overall landscape.

While the full potential for integrating functions of wastewater renovation, wildlife habitats, recreational open space, artworks, interpretive trails, educational centers, and visually attractive landscapes has not yet really been explored in many completed projects, there are a number of examples from various parts of the country that may be useful in conveying visions of the inherent possibilities in other projects. This book has presented examples of constructed wetlands projects incorporating residential landscape features, wildlife habitat, open space, interpretive signage, artwork, and other features. It can be fairly stated that virtually every constructed wetland installation, of any scale, provides increased wildlife habitat, even at a backyard level. Some of the large examples of constructed wetland features that are part of a wastewater treatment system but which also provide significant wildlife habitat and recreational open space are the Arcata, California, system; the Mountain View Sanitary District in California; the Showlow Pintail Lake system in Arizona; and the Columbia, Missouri, systems, all of which are described later in this chapter.

One of the only constructed wetland wastewater treatment systems in the world in which the treated wastewater is collected in an aquatic study pond and utilized on a systematic basis in the teaching curriculum was designed by the authors for the Los Padillas Elementary School in the South Valley area of Albuquereque, New Mexico. This system is described later in this chapter.

Projects in which artists were centrally involved in helping to convey information, both explicitly and subliminally, on biological processes in nature were described in Chapter 9. The present an exciting challenge to engineers, landscape architects, government officials, and others in working to provide more richness within public infrastructure projects.

As the understanding of constructed wetlands and their potential multiple uses develops in each region of the country, and on a worldwide basis, we will see more and more innovative examples of these systems, in many permutations, integrated into park and recreational facilities, golf courses, resorts and ecolodges, school grounds, housing developments, botanic gardens, and other facilities. There will be more recognition of the importance to people of visual contact with wildlife, and more senstitivity to incorporating features and vegetation of benefit to a variety of wildlife. The unique educational and interpretive values inherent in constructed wetlands will also receive greater recognition as essential to recognize and develop in any publicly accessible facility. The melding of biology, landscape architecture, engineering, and art can produce memorable concepts that have the ability to connect people with their environment and to convey information in both a direct and a subtle manner, This connection between disciplines—science, art, engineering, design—and between humans and their surroundings is crucial, and must be nourished and supported in every appropriate venue to the greatest extent possible.

Most projects dealing with wastewater or stormwater have been controlled by public works department personnel, who have defined their scope in narrow terms limited by both budget considerations and their own lack of understanding of alternatives. As stormwater and wastewater can both be quite unpredicable in quantity and quality—whether at a residential or community scale—it is not surprising that regulators and engineering firms in general take a very conservative cookbook approach to the design of treatment facilities. The first priority, as quite properly viewed by most engineers, is to choose equipment and designs that they are familiar with from past experience and that will operate with a fair degree of reliability over a wide range of changing conditions. Innovative systems are viewed as risky, and many engineering firms are unwilling to stake their reputation on any systems with which they are not familiar. Thanks to a great deal of positive publicity, along with professional conferences, papers, and reviews of existing systems by the U.S. EPA, the Water Environment Federation, and some local regulatory agencies, this situation is rapidly changing. It is much more common now for the majority of engineering firms dealing with wastewater issues to consider constructed wetlands as a feasible alternative for many projects.

Stormwater wetlands are presently not as widely accepted, both because there has been less technical information and monitoring of those systems and because of the higher level of maintenance they require. While there is great potential to influence some projects that have already been budgeted and narrowly defined as "stormwater engineering" or "wastewater engineering," this usually entails citizen-inspired efforts bolstered by environmental groups interested in broader issues who can transform such projects into ones with multiple purposes.

There is much greater potential, both within government agencies and within community groups and environmental organizations, to develop long-range plans based upon sustainable, multiple-use principles that are woven into public law and policies at the local level. There are many examples of this symbiotic effort, nurtured by the shift in public attitudes regarding natural systems, between citizens, organizations, and government officials in developing creative and sustainable solutions to pollution and stormwater problems in the environment. There is no question, however, that without the presence of an enthusiastic and effective advocate within the system—that is, within government agencies—the task is much more difficult. Throughout the country, such individuals have had long-lasting positive effects in promoting a new attitude toward projects that formerly were pigeonholed into a narrow category not worthy of public involvement or contribution by landscape architects, artists, or biologists.

This chapter provides a brief description of some of the more creative applications of constructed wetlands in the landscape, both for wastewater and stormwater treatment, along with various multiple-use functions ranging from wildlife habitat and trails to educational and recreation. The first project presented is still in the planning stage for the Stapleton Airport site in Denver.

STAPLETON AIRPORT REDEVELOPMENT PLAN

Designers: Anndropogon Associates, Inc.; William Wenk Associocates

Stapleton Airport finally closed in 1995 after several delays in opening the new Denver International Airport. The Stapleton Redevelopment Foundation, a private nonprofit group working with the City of Denver, sponsored a plan to create a new community on the 4,700 acre site that would be based from the beginnning on sustainable principles (Figure 10.1 and 10.2). A core team of consultants was assembled, including the innovative firms Andropogon Associates, Ltd., of Philadelphia and William Wenk Associates, landscape architects, of Denver. Both firms have a history of developing plans for stormwater management that combine natural drainage, infiltration, and vegetative solutions within creatively designed landscapes. A detailed site analysis was prepared for the Stapleton site, and natural habitats in the surrounding areas were identified for integration into the plan. A major central open space was identified for the redevelopment, which will represent the biggest single addition to the Denver parks system in fifty years. The stormwater management system utilizes riparian habitats with indigenous plantings that form the spine of the open space system and are fully integrated with the recreational facilities. Andropogon Associates developed the concept of habitat restoration within the project to provide contact with indigenous plant and animal communities of the western high plains. The plans call for reintroduction of native habitats at all scales, beginning at the garden or schoolyard level and continuing to a regional-scale reestablishment of sandhills prairie and restoration of the historic forested stream channels of Sand and Westerly creeks.

Plans call for meandering vegetated swales, wetlands, and sediment ponds to handle stormwater runoff and minimize pipes and hard structures. On-site infiltration of runoff is a major goal, and even though this approach obviously requires

Figure 10.1 Stapleton Airport redevelopment plan (William Wenk).

Figure 10.2 *Stapleton Airport redevelopment plan (William Wenk).*

more land than would a subsurface piped stormwater drainage system, the land used to manage and convey stormwater in this plan also becomes an important recreational and open-space amenity for the surrounding community while providing wildlife habitat.

ARCATA WASTEWATER MARCH AND WILDLIFE SANCTUARY

Designer: Frank Klopp and Robert Gearhart

The Arcata, California wastewater marsh project, now officially designated the Arcata Marsh and Wildlife Sanctuary (AMWS), is probably the most widely publicized constructed wetland facility in the world, having been featured in *Smithsonian, Audobon, Time,* and many other magazines and journals, both popular and technical. As an exemplary demonstration of the multiple-use potential of constructed wetlands, the Arcata facility combines wastewater renovation, wildlife habitat, educational and interpretive features, and public recreational access within its 150 acres. As is often the case with the most innovative and successful projects, several key individuals with vision and patience provided inspiration and guidance that helped make this facility a reality. Robert A. Gearheart, Professor of Environmental Resources Engineering at Humboldt State University in Arcata, was the father of the idea of the Arcata system. The AMWS is essentially a tertiary unit that was added to the existing 2.6-million gallon per day primary and secondary wastewater treatment facility between 1985 and 1989.

Over 5 miles of foot trails are open for public access, and over 130,000 people a year use the site for recreation and education. The Redwood Regional Audobon Society sponsors weekly nature walks at AMWS, conducted by trained docents, which draw approximately 900 people a year. The area is widely used by Humboldt

State University and local public schools for field trips and various educational programs, and even for research and graduate courses in wetland management. The various interpretive programs are seen as having contributed greatly to public awareness of the multiple benefits of the system. Table 10.1 the wide variety of visitors and uses typical of the Arcata facility, along with proposed measures to deal with the impacts of such heavy use.

COLUMBIA WETLANDS WASTEWATER TREATMENT PROJECT

Designer: Metcalf & Eddy, Inc.

This project is an excellent example of the power of citizens and environmental groups taking the initiative to influence the decisions made by public officials responsible for wastewater managment. In 1987, the City of Columbia, Missouri, faced a common problem: a treatment plant reaching capacity at about the same time that more stringent water quality standards were being applied. Even though the initial engineering studies identified an expansion of the existing plant and discharge to the Missouri River as the most economical solution, the citizens and local environmental groups insisted on evaluating natural approaches to solving the problem while attaining even higher treatment standards. In a truly admirable co-

TABLE 10.1 Gross Use Estimates, Negative and Noncompatible Activities, and Methods of Resolving Negative Impacts of Recreational Activities Observed in Arcata Marsh and Wildlife Sanctuary

Recreational Use	Estimated Use	Management & Mitigation Interventions
Picnicing & relaxing	75–175 people a day; 15–45 min. per visit	Traffic control, parking, placement of picnic tables and refuse cans; volunteer cleanup program; integrated mosquito control programs
Birdwatching and nature studies	5–50 people a day; 2–4 hours per visit	Traffic control, parking, designated areas with blinds, to enhance the experiment and to allow all-weather usage; integrated mosquito control program; docent program by Audubon Society
Jogging and walking	20–80 people a day;	Designated trails, all weather surface-chips and bark; integrated mosquito control program
Education	50–200 people per month; 2–4 hrs. per day	
Fishing	Light and seasonal	Borrow pit next to south lake and fishing jetties designated to direct bank fishing users away from critical bird areas; integrated mosquito control probram; public education to values of trophy fishing program and values
Photo/art	30–75 people per month; 2–4 hrs. per visit	None

Source: Gearhart and Higley (1993, p. 567).

operative effort between many public agencies, citizens, and design firms, an innovative plan was created to develop three wetland treatment units that would then be used to supply treated wastewater to a newly restored riverine wetland along the Missouri River managed by the Missouri Department of Conservation. The wetlands consist of 1,300 acres within the 3,656-acre Eagle Bluffs Wildlife Area. The entire area had previously been intensively farmed for many years; and this project represented an attempt to restore some of Missouri's presettlement wetlands, 90 percent of which had been lost.

Three wetland treatment units totaling approximately 91 acres were constructed to treat an average flow of 17.68 million gallons per day. The wetlands were sized to produce an effluent meeting National Pollutant Discharge Elimination System (NPDES) permit limitations of 30 mg/L BOD and 30 mg/L suspended solids. As the existing wastewater plant receives a significant quantity of stormwater flow, the wetland components had to be sized to permit 2.63 m^3 per second (60 million gallons per day) of flow to pass through the wetlands. A complex system of levees, pumps, channels, and ponds make up the Eagle Bluffs Wildlife Area and provide controls capable of mixing river water and treated wastewater; diverting water for temporary storage or emergency diversion; and utilizing both gravity flow and lift pumps. This project entailed cooperative efforts between the City of Columbia, the Missouri Department of Conservation, the U.S. Geological Survey, the U.S. EPA, the Army Corps of Engineers, the U.S. Soil Conservation Service, the U.S. Fish and Wildlife Service, and the University of Missouri.

The wetlands are designed to encourage public use, with a river outlook, trails, restrooms, a visitor's center, wetland boardwalks, scenic platforms, photograph blinds, self-guided nature trails and auto tours, and other public features either built or in the planning stage. Local officials estimate that at least 65 percent of the people visiting the area will be interested in walking, hiking, birding, mushroom gathering, wildlife observation, photography, and scheduled educational programs. Traditional uses such as hunting, trapping, and fishing are also allowed during certain times of the year. In addition, special public use events such as Eagle Days, a Spring Waterfowl Viewing Weekend, and a Shorebird Migration Weekend are in the planning stage.

ORLANDO EASTERLY WETLANDS RECLAMATION AND PARK

Designer: Post, Buckley, Schuh & Jernigan

At a time of unprecedented population growth in Orlando, Florida's, Easterly Service Area, which includes portions of Orlando, other communities, and unincorporated areas of two counties, a building moratorium was being considered due to lack of an acceptable effluent disposal method. The Iron Bridge treatment facility had almost reached its design capacity of 24 million galllons per day and needed expansion to 40 million gallons per day to accommodate projected growth. Post, Buckley, Schuh & Jernigan, Inc. (PBS&J), was asked by the City of Orlando to perform a comprehensive study of unconventional alternatives and recommend a permittable, environmentally acceptable solution. The original concept for the project came from Phil Searcy, P.E., executive vice president of PBS&J, the consultants to the City of Orlando for the project. The original notion was for a chain of lakes to purify effluent, and project director Phil Feeney, P.E., and project engineer JoAnn Jackson, P.E. then developed the final concept in detail (Figure 10.3).

The Orlando Easterly Wetlands is a 1,200-acre surface flow constructed wetlands facility designed to remove nitrogen and phosphorus from 16 million gallons per day of advanced wastewater treatment effluent from the Iron Bridge Wastewater Treatment Facility prior to discharge to the St. Johns River in central Florida. The project reclaimed 1,200 acres of historic wetlands that had been drained and converted to pasture. The system includes cattail-bulrush, grass/herbaceous, and forested wetland communites, and represents the first large-scale constructed wetland designed from the beginning both to treat wastewater and to provide wildlife habitat and interpretive trails. Opened in 1987, the wetlands system provides habitat for over 150 plants, 140 bird species, and numerous animal species, including otters, foxes, deer, turtles, snakes, and alligators. A number of rare and endangered species

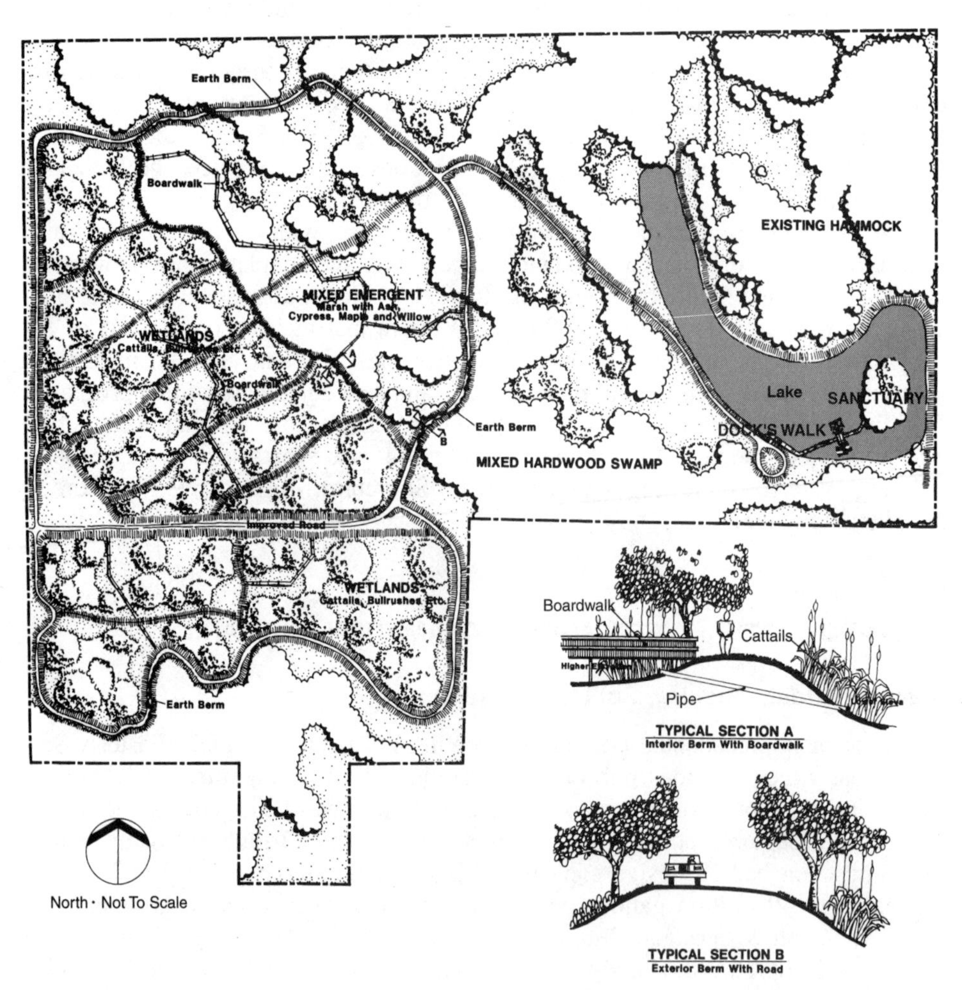

Figure 10.3 *Ironbridge, Florida constructed wetlands: schematic plan (Post Buckley Schuh & Jernigan).*

use the Orlando wetland system, including the Everglades kite, bald eagle, wood stork, sandhill crane, and eastern indigo snake. Since 1987, this wetland treatment system has consistendly reduced average annual influent total phosphorus and nitrogen concentrations to less than 0.095 mg/L and 0.905 mg/L, respectively. The wetlands were created with twenty-three water control structures and 18 miles of berms creating 17 wetland cells, to allow management of flow patterns and depths. A total of 2.1 million aquatic plants, including 120,000 trees, were planted to create three distinct plant communities: a deep marsh, a mixed marsh, and a 400-acre hardwood swamp. Mixed understory marsh plants were planted in the hardwood swamp in addition to the trees, and a 120-acre lake was created from excavation in the hardwood swamp that provided fill for the berms. Water meanders through the complete system in about thirty days. The first stage of nutrient removal is provided by the cattails and bulrush in the deep marsh; the mixed marsh provides polishiing of the remaining nutrients; and the hardwood swamp serves as a final nutrient sink. PBS&J was retained by the City of Orlando to manage and operate the wetland system. An 80-acre lake was created to provide fill for the berms and is now a central feature of the facility. The portion of the Iron Bridge Easterly Wetlands that is open to the public is called Orlando Wetlands Park and features a 2-mile birding route that encircles the prime wildlife and bird habitat. There are also hiking trails, biking trails, and interpretive guides, which create a unique feature within one of the fastest-growing areas of the country (Figure 10.4). The area is used as a natural laboratory by area schools for environmental education programs and as a research facility by students and professors of the University of Central Florida. Park usage is monitored by a sign-in sheet for visitors, who are urged to record their wildlife sightings for the record. The best birding locations are identified on the trail guide.

It is important to point out that innovative projects such as this one usually require a client who supports the concept and is willing to fight for it, as did the

Figure 10.4 *Ironbridge wetlands (Robert Knight, CH2MHill).*

City of Orlando. The project required approximately twenty-four permits from seven separate regulatory agencies. In the end, there was virtually unanimous agreement that the Easterly Wetlands Reclamation Project was a resounding success. The system met Orlando's time requirements and prevented the establishment of building moratoriums; its cost was far less than that of conventional disposal methods; and the life-cycle cost of the wetlands was 15–30 percent less than that of conventional treatment and disposal alternatives. According to Alex Alexander, P.E., District Manager for the Florida Department of Environmental Regulation, this is a "really beautiful project." It is also a project that has received numerous awards, both from regulatory agencies and from enviromental and engineering organizations—a harbinger of things to come, one would hope!

SACRAMENTO COUNTY DEMONSTRATION WETLANDS

Designer: Nolte and Associates

In an effort to determine the ability of constructed wetlands to treat secondary effluent to meet the stringent water quality standards of the California Inland Surface Water Plan, Sacramento County constructed a 22-acre long-term demonstration project adjacent to its 181 million-gallon-per-day regional wastewater treatment plant. The facility was designed to include a series of eleven wetland cells, each 50 feet wide and 1,260 feet long, constructed with deep-water areas at the ends and intermediate points to provide a refuge for gambusia (mosquito fish). The system was designed to receive about 1 million gallons per day of disinfected secondary effluent from the Sacramento Regional Wastewater Treatment Plant (SRWTP). Effluent from the treatment cells is routed through a 2-acre wildlife habitat wetland before being returned to the (SRWTP) influent sewer system. The effluent will continue to be discharged to the sewer system until the project is able to demonstrate adequate performance and an NPDES permit is received to allow discharge to the natural wetlands along Laguna Creek. Among the contaminants the project will monitor are heavy metals, ammonia, and trace organic compounds. The habitat wetlands will provide a bioassay to determine the feasibility and effects of the discharge into a natural wetland system, and will have two islands for waterfowl nesting along with areas of deep open water. One of the interesting aspects of the development of the wetland cells is that the rapid development of the aquatic plants appeared to derive more from seed contained in the natural wetland soil than from the bulrush and tule tubers that were planted.

SHOW LOW WETLANDS

Designer: Mel Wilhelm

In the high country of northeastern Arizona, a constructed wetlands system receiving wastewater from aerated lagoons, developed by the City of Show Low, on National Forest Service lands, has developed into the best waterfowl breeding habitat in the state. The first wetland was created in 1970 by pumping effluent 2 miles to a 45-acre natural depression known as Telephone Lake. The next wetland development was undertaken by the U.S. Forest Service in 1978 through the development of other natural depressions into marshes and lakes, which included wildlife islands to enhance waterfowl reproduction. These features became known as Pintail Lake (57 acres) and South Marsh (19 acres). By 1982, population increases exceeded the treatment plant's capacity, and the quality of the effluent delivered to the marsh-lake system was poorer. Working closely with the U.S. Forest Service, the Show Low developed a long-term improvement plan that included improving the existing aerated lagoons, adding stabilization ponds, and adding additional marsh capacity for final treatment and reuse. In 1985, a major expansion of the wetlands was undertake with the construction of the 49-acre Redhead Marsh. At that time, the upgraded system was designed to handle 1.4 million gallons per day of wastewater to service a population of 13,500 to the year 2006. Because the wetland system was designed as an integrated complex of lakes and marshes, the objectives from the beginning included major considerations of wildlife habitat. Many other constructed wetlands around the country have inadvertently become wildlife habitats, some more successfully than others, but the fact that individuals such as Mel Wilhelm, formerly with the U.S. Forest Service in the Show Low area, were centrally involved in the design and construction of this system ensured its success as a multiple-function facility. Mr. Wilhelm worked closely with the community and rallied a high level of volunteer effort in planting wetland and wildlife food plants such as Japanese millet in and around the wetlands.

The entire system was designed to allow considerable flexibility in controlling water levels in each of the marshes and lakes. Water control structures provide for diversion of water away from some ponds to allow them to dry up, if necessary, to accomplish vegetation management goals. In addition to the original plantings of sago pondweed and hardstem, softstem, and alkali bulrush, other plant species such as cattail, water grass, spike rush, and sedges have become established naturally in the wetland system. The creation of water sites within an otherwise arid landscape in this part of Arizona has exceeded the original expectations and has resulted in one of the most attractive artificial wetland complexes in the country. As a wildlife habitat, the Show Low system is superb, and has attracted nesting colonies of cormorant and black-crowned night herons, along with at least ten bird species classified at endangered. A 1991 survey documenting bird use found 120 different species of birds using the created wetlands. Animals such as elk, mule deer, pronghorn, black bear, coyote, raccoon, and others also have been observed in the area of the wetlands.

This project represents a cooperative effort between the City of Show Low, the Arizona Game and Fish Department, and the U.S. Forest Service. It has been supported by local organizations and schools, which regularly use the facility for environmental field trips. The success of the installation surpassed all expectations, and the site has become the most productive waterfowl breeding area in the State of Arizona.

EL MALPAIS NATIONAL MONUMENT: MULTIAGENCY WELCOME CENTER

Engineers and Architects: Wilson & Company; Solar Architects: Mazria Associates; Landscape Architects: Campbell Okuma Perkins Associates, Inc.; Wetlands Engineers: Southwest Wetlands Group

As part of an effort by a multidisciplinary team of architects, landscape architects, and engineers to develop plans for a new visitor's center to convey information about a fascinating region in and adjacent to one of the nation's most recent national monuments, a series of value analysis studies were conducted on handling of the center's wastewater. Although a connection to the nearest sewer line from the City of Grants, New Mexico, was about 1 mile away, every other potential alternative was considered and evaluated. Grants had an agreement with the neighboring Indian pueblos not to discharge any effluent into groundwater, a condition known as a *zero discharge* requirement. The team engaged by the National Park Service included engineering and landscape architectural firms experienced in constructed wetlands design, and this obviously gave impetus to the effort to explore the feasibility of constructed wetlands as one method of treating wastewater. The consultant team had pointed out the existence of a number of unique naturally occuring wetlands within the basaltic lava flow area close to the national monument. They believed that a constructed wetland could be a natural-appearing enhancement to the enjoyment of the site, as well as an educational tool regarding the natural method of treating the wastewater.

Five representatives from the Denver Service Center of the National Park Service, along with six represenitives from the consultant team, spent a day brainstorming and evaluating six alternatives for treating wastewater within a *value engineering* format. The constructed wetlands alternative called for capturing roof and parking lot runoff and treating it in wetland cells along with the building wastewater storing the treated water in ponds, and recycling it to the facility for toilet flushing and landscape irrigation. The alternative of hooking into the Grants sewer lines would have required a lift station feeding into a force main, acquiring easements, blasting through basalt, and so on.

The criteria and weighting which were developed by the team were as follows:

Criteria	Value
A. Low maintenance and operating costs	8.2
B. Low initial cost	4.8
C. Interpretive value	7.6
D. Low end-use energy consumption (resource depletion consideration)	10.1
E. Relation to sustainable design criteria (excludes energy consumption)	
F. Low environmental impact	10.6
G. Aesthetics—visual, odor, noise, etc.	7.6
H. Reliability and ease of operation	8.2
I. Potential to accommodate expansion of facility	3.1

Value: 1 = low weight value; 10 = high weight value.

Each team member assigned his own value to each criterion, and then the values were averaged. The weighted criteria were then applied to each of the six alternatives for wastewater treatment. The final result was that the constructed wetlands alternative was the clear winner, with by far the most points.

The overwhelming judgement of the team was that the constructed wetlands alternative would create a strong interpretive element with high educational value. It was also the alternative most consistent with a sustainable design initiative, a central organizing concept adopted by the National Park Service. If any federal government agency is in a postion to take the lead in sustainable, low-impact design and development, it is the National Park Service, which operates facilities in some of the most sensitive areas of the country

LOS PADILLAS ELEMENTARY SCHOOL

Landscape Architects/Planners: Campbell Okuma Perkins Associates, Inc.; Project Director: Craig Campbell; Wetlands Engineers: Southwest Wetlands Group, Inc.

The Outdoor Classroom and Constructed Wetlands Project for Los Padillas Elementary School in Albuquerque, New Mexico, represents the first utilization of wastewater for both organized educational and wildlife habitat purposes in the Southwest. Faced with a failing sand mound wastewater disposal system within a site with very high groundwater, the school became interested in the potential integration of a constructed wetland into a special 4-acre site adjacent to the school that only recently had been discovered to be owned by the school district.

As is often the case, one enthusiastic and dedicated person within the school system led a long effort to gain financial support for a demonstration project on the site, which had already been utilized by the school as an ecology study area and wildlife sanctuary. Funding of about $200,000 was finally obtained from a combination of State of New Mexico agencies and the Albuquerque Public Schools administration. It was earmarked for a demonstration constructed wetland to replace the sand mounds, an outdoor classroom shelter for up to fifty students, and development of educational mini-botanic garden exhibits.

After several sessions with the school's Ecology Committee and community representatives, the final plans for the site were developed. Included in the project, in addition to the shelter and the constructed wetlands, were an aquatic study pond, a nature study trail, plant community exhibits, stone seating, and an interpretive brochure. In response to a request by the students, an adobe arched gateway was also designed for the project and a stone sign embedded in the arch that identified the site as the Los Padillas Wildlife Sanctuary.

While it would have been advantageous to utilize gravity flow for the entire system, the flatness of the entire site mandated a series of pumps starting at the first septic tank and ending with a sump pump forcing the effluent past a UV disinfection lamp into the pond. Such installations are problematic, as they require regular maintenance to maintain the clarity of the lens through which the UV radiation must be transmitted. In addition, when installed in an outdoor location within a vault, they are subjected to extremes of temperature that may affect their performance. Other options for disinfecting wastewater that may be exposed to public contact are combination ozone and UV systems, chlorination, filters and brominators, and other methods, each with advantages and disadvantages for each condition.

Among the plantings installed in the aquatic study pond were cattail, softstem bulrush, yellow iris, pennywort, water lilies, and others. In addition, Elodea were installed on the bottom of the pond as good aeration plants; however, the initial cloudiness of the water prevented these plantings from becoming established. Fil-

Figure 10.5 *Students involved in wetland planting, Los Padillas Elementary School (Craig Campbell).*

amentous algae presented a real problem in the first year of operation, due in part to the fact that the wetlands had not been established and excess phosphorous was being discharged into the pond. The addition of duckweed (*Lemna* spp.) and the regular removal of algae resulted in the almost total disappearance of the algae by summer's end and the dominance of duckweed, which, along with the water lily pads, tended to shade the water sufficiently to prevent the growth of algae. The long-term objective is to achieve almost total cover of the pond surface by water lily pads to minimize algae problems (Figures 10.5, 10.6, 10.7, 10.8).

Figure 10.6 *Los Padillas Elementary School (Craig Campbell).*

Figure 10.7 Los Padillas Elementary School (Craig Campbell).

Among the issues confronting the designers of the Los Padillas system were liability related to control of the site and the pond and the necessity to provide disinfection of the wastewater prior to its discharge into the open pond. The first issue, that of liability, was addressed by installing a complete enclosure around the site with a 5-foot fence and only one access point. The school controls and maintains the access through a normally locked gate at the new adobe arch.

The project has been visited by many educators and administrators from all over the country, as well as from Japan. They have been impressed with this creative

Figure 10.8 Los Padillas Elementary School (Craig Campbell).

and attractive example of reuse and integration into an educational program of an otherwise wasted resource. The pond has become a major attraction, featuring frogs, water boatmen, dragonflies, and many other aquatic inhabitants, as well as periodic visits by ducks.

BINFORD LAKE/BUTLER CREEK GREENWAY MASTER PLAN

Designers: Walker & Macy

In 1992, Walker & Macy was commissioned to develop a master plan for 4 miles of greenway along Butler Creek and around Binford Lake for the City of Gresham, Oregon. The plan was based on a holistic approach that balanced water quality, wildlife, recreational, and educational concerns while providing a prototype for stormwater management and aesthetic enjoyment for the city. It incorporated bio-filtration swales planted with specific grasses to intercept and filter stormwater runoff before it enters the lake; a boardwalk with interpretive signage; an enhanced wetland to filter pollutants and provide forage and shelter for wildlife; and a water quality pond to intercept stormwater fully designed with appropriate plantings for a water level ranging from 6 inches to 2 feet. A positive statement was made by the interpretive signage which not only described the project goals and problem areas addressed by the plan, but also cautioned users about fertilizers and toxins and conveyed the concept of integrated pest management, along with the biological means of addressing water pollution.

JACKSON BOTTOM RESOURCE MANAGEMENT MASTER PLAN

Designers: Walker & Macy

In 1988, Walker & Macy was retained by the City of Hillsboro, Oregon, and the Unified Sewerage Agency to develop a master plan for Jackson Bottom, a lowland area within the floodplain of the Tualatin River. Central to the plan was an analysis of the area's suitability for a biological process of sewage filtration through wetlands (Figure 10.9). The 3,000-acre project was one of the first in the Northwest to explore low-cost, environmentally sound means of processing wastewater to comply with state water quality standards while at the same time recognizing the importance of expanding the wildlife population, preserving open space, and providing recreational opportunities. A trail system with observation blinds and an interpretive center are planned to promote environmental education. Significantly, Robert Gearheart—the father of the Arcata, California, AMWS,—was a consultant on the project.

MT. VIEW SANITARY DISTRICT WETLANDS

Designer: Richard Bogaert

The Mt. View Sanitary District (MVSD) Wetlands in Contra Costa County, California, represent one of the first deliberate efforts in the country to make beneficial

Figure 10.9 *Jackson Bottom, Oregon* (*Walker & Macy*).

use of secondarily treated wastewater for creation of wildlife habitat and an educational facility. Originally constructed in 1974, the wetlands were designed to receive an average 1.4 million gallons per day of treated wastewater. The wetlands were doubled in size in 1977 and doubled again in 1984. Among the improvements and modifications made recently to the facility are a full-scale UV disinfection unit to replace chlorination, which had presented various problems, including excess ammonia. In addition, a portion of the wetlands was drained and deepened, with new wildlife habitat islands installed.

The decision to develop wetlands along with an ongoing management program grew out of stricter discharge standards adopted by the State Water Resources Control Board and the Regional Water Quality Control Board (RWQB) in the 1970s. By encouraging wastewater reclamation, and by allowing exceptions where environmental benefits could be demonstrated by the discharge, the RWQB recognized the multiple advantages of wetland creation to offset the huge historic loss of wetlands in California. In 1977 the RWQB adopted "Policy Guidelines on the Use of Wastewater to Create, Restore, Maintain, and/or Enhance Marshlands," providing dischargers with clear guidelines to follow when seeking exceptions from the usual discharge dilution requirements.

The MVSD undertook from the beginning a sound management program designed to enhance the wetlands' wildlife habitat and to provide information for the public and the scientific community. A biologist put on staff early in the project was the key individual in overseeing both wetlands management and data collection, as well as coordinating visits by scientists, school classes, and others. A local college schedules regular field trips for its biology classes, and the facility receives many visits by representatives of other communities and professionals interested in the design, development, and performance of wetlands. The largest category of visitors, as appears to be typical of other similar facilities, consists of birdwatchers; educational groups are a close second.

The MVSD wetlands provide an exceptionally rich variety of wildlife, ranging from deer, beaver, river otters, skunk, raccoon, and fox to over 123 bird species, many of which nest within the site. The local Audubon Society supports and regularly uses the facility. The district encourages public use of the wetlands, which is open from 8:00 A.M. to 4:30 P.M., and provides an interpretive guide who identifies a series of stations where particular features may be observed. Visitors are required to sign in and out. This provides a good record of visitation, which indicates that birdwatching is the most popular activity.

The newer forest/marsh unit is used primarily as an evapotranspiration system, which operates year round in the mild Mediterranean climate of the Bay area. Evergreen trees such as redwood, eucalyptus, and Monterey pines have been planted and utilize water from the system year round. The philosophy underlying the operation of the MVSD wetlands is best understood by quoting from its status report of 1984:

> . . . [T]he very existence of the wetlands relies on their proper management to produce continued net environmental benefits. . . . Wetlands development creates non-contact water recreation, aesthetic enjoyment, and enhances wildlife habitat. The State recognized these beneficial uses of wetlands. . . . It is estimated that 90% of the wetlands within the San Francisco Bay region have been drained or filled to make way for agricultural, residential, and industrial uses. . . . This loss of wetland habitat has resulted in greatly reduced wildlife and migratory waterfowl populations. . . . MVSD has intervened by creating a viable wetlands that has supported many generation of aquatic invertebrates, waterfowl, and other wildlife. Biologists have identified 123 species of birds, 69 species of plants, 26 species of animals, and 34 species of aquatic invertebrates in the wetlands. The wetlands are a successful nesting area for ducks and other waterfowl. They also serve as a wintering, feeding, or resting spot for migratory birds on the Pacific Flyway. . . . [T]he scenic beauty of the wetlands . . . invites many motorists to enjoy the reflection of cattails on the water, migrating ducks taking wing, or a lone snowy egret wading in the marsh. Aesthetically the wetlands are an oasis in a desert of industrial facilities. MVSD believes that aesthetic enjoyment can be increased through education, and therefore makes the wetlands available to school groups, community groups, and individuals who are interested in learning more about ecology, bird life, or other marsh phenomena. As educational opportunities are expanded to reach more people, the wetlands increase in value to them. . . . MVSD has developed and gained the enthusiastic support of many locals, state professionals, and even a few birdwatchers from around the world. The District continues to extend its welcome to birdwatchers, educational groups, and others to experience the educational and recreational opportunities available in the enhanced wetlands.

This statement represents a truly admirable commitment to management of wetlands for multiple uses, by a relatively small agency with a total staff of ten, whose primary concern is the treatment of wastewater; one hopes its will influence other administrators and government officials around the country and continue to stimulate the imagination of all visitors. It is no accident that some of the strongest support for constructed wetlands for wastewater treatment comes from Audubon Society chapters, Ducks Unlimited, and other groups interested in expanding wildlife habitat.

MITCHELL LAKE WETLANDS AND CHAVANEAUX GARDENS

Designer: San Antonio Water System

The City of San Antonio has recently begun developing a wetlands and wildlife habitat on the site of a natural lake that was long utilized as a sludge management facility. The lake had originally been used as the first sewage treatment facility and to provide irrigation water to adjacent farmlands from the late 1800s to the 1930s. A 589-acre Chavaneaux Gardens property north of the lake was also utilized in the sludge management operation by land application. A change in sludge management techniques made these operations unnecessary, and in 1987 the Mitchell Lake Recovery Advisory Subcommittee was created to develop a comprehensive recovery plan for the lake. The plan, completed in 1989, called for development of a wildlife refuge and creation of a permanent water cap in the form of constructed wetlands over the sediments in the basins previously used for sludge management. The construction of the wetlands began in fall of 1993 and was completed in November 1994. Mitchell Lake will provide treated effluent to the constructed wetlands, from which it will be circulated back to the lake. The lake will provide effluent for irrigating a golf course and secondarily will supply water to the San Antonio River Authority tunnel, which provides water to the San Antonio River for irrigation of other areas.

Long-range plans for the Mitchell Lake wetlands system include creating educational and environmental activities with the area universities and school systems; promoting birdwatching and other activities; and developing amenities such as restrooms, shelters, benches at the wetlands, and educational center, trails, and other improvements. Although the initial creation of the wetlands was undertaken by the San Antonio Water System, the possibility of acquisition by the City of San Antonio Parks Department is being considered for eventual permanent operation and management. A public/private partnership, the Mitchell Lake Wetlands Society, has been established to support ongoing operation of the facility by residents of the city.

BEAUMONT CONSTRUCTED WETLAND

Designer: Schaumberg and Polk

The Water Utilities Department of the City of Beaumont, Texas, built a constructed wetland as the final polishing phase of their wastewater treatment system. When it was originallly designed, the facility was not envisioned as a mulitple-function constructed wetland; it was primarly seen as a functional element of the wasterwater treatment system. Facing the familiar choice of either expanding the existing wastewater treatment facility or piping the treated wastewater over a long distance to the Neches River, both of which would have been costly options, the city researched the use of constructed wetlands in other parts of the country and decided that this was the most cost-effective alternative to meet the more stringent U.S. EPA requirements. The constructed wetland, named Cattail Marsh, provides additional water to enhance a preexisting 250-acre natural wetland at Hillebrandt Bayou. Prior to the installation of the constructed wetland, the city had a 20 mg/L BOD_5 and 20 mg/L TSS permit that allowed it to discharge into the Hillebrandt Bayou. With the more stringent streamwater quality standards of the U.S. EPA, the city could not continue discharging to the bayou without meeting the new limits of 3 mg/L ammonia and a BOD_5 limit of 5 mg/L. If the more costly alternatives were undertaken, removing the continuous flow of water to the Hillebrandt Bayou would probably have had serious consequences. Therefore, the constructed wetland alternative, at about half the cost of expanding the existing treatment plant ($18 million), was selected as not only the most cost-effective alternative, but one that could provide multiple benefits such as a wildlife refuge, a stormwater retention area, and a research base for the universities.

This 650-acre wetland, which began operation on September 3, 1993, joins the MVSD wetland and the Arcata marsh system as among the largest in the country. The wetland is divided into eight cells originally planted with 270,000 California bulrush (*Scirpus californicus*), along with arrowhead, pickerel weed, and smartweed. Other natural aquatic vegetation such as cattails, rushes, coontail, and duckweed have also become established in the wetland. The bulrush, as the primary plant was selected due to its effectiveness in ammonia removal and its ability to tolerate water depths of up to 3 feet. The wetland system can be operated in series or in parallel, and the water can be maintained at levels ranging from several inches to several feet, providing optimum control. The design also provides alternating shallow cells, deep cells, and then shallow cells, which aids in nitrification and denitrification. The state and federal permitting agencies have allowed the following permit limits for this facility:

$CBOD_5$ 10 mg/L
TSS 15 mg/L
DO (min.) 5 mg/L
Ammonia 3 mg/L

After only one year in operation, this system achieved ammonia nitrogen removal efficiency of 81–99 percent and has allowed the City of Beaumont to meet its permit limits ever since it began operating the wetland system. The $CBOD_5$ has consistently been less than 2 mg/L, TSS less than 10 mg/L, and ammonia levels frequently less than 0.1 mg/L.

The Cattail Marsh has become a tourist attraction, as well as a major wildlife habitat and recreational feature for local residents. In recognition of the multiple functions of the facility, the Water Reclamation Division has added a biologist to their staff who is actively engaged in monitoring wildlife use of the facility and developing interpretive functions for the public. The wetlands are fortuitously located adjacent to an existing park—Tyrrell Park—and are now imaginatively presented as an integrated unit, Tyrell Park and Cattail Marsh, with greatly expanded hiking, birdwatching, wildlife photography, and other options. The park itself offers recreational vehicle facilities for overnight campers, picnic areas, a children's playground, an eighteen-hole golf course, a garden center area, and riding stables. The range of activities associated with both the park and the adjacent marsh creates one of the richest and most varied public recreational and wildlife habitat facilities anywhere in the country. Over 350 bird species, including egrets, pelicans, roseate spoonbills, ibis, and ducks, use the wetlands annually. In addition, of course, many species of fish, reptiles and amphibians, and mammals inhabit the wetlands. Given the location of this particular facility, however, the city recognizes that not all wildlife is necessarily benign; visitors are warned to watch out for poisonous snakes, fire ants, and alligators. There are 8 miles of graveled levee maintenance roads that are available for hiking, jogging, biking, horseback riding, birdwatching, picnicking, and other activities.

IN SUMMARY

There are many additional interesting applications of constructed wetlands around the world that are not addressed in this volume. Ecolodges in Costa Rica, schools and campgrounds in Europe, a resort in India—the list is endless and presents fascinating possibilities for future research and cooperative efforts in sharing the knowledge produced in this evolving field—one small but important effort in the worldwide move toward sustainable development. In closing, a prescient quote from Michael Hough (1990) seems appropriate:

> [T]he principle of investment in nature, where change and technological development are seen as positive forces to sustain and enhance the environment, must be the basis for an environmental design philosophy. Its principles of energy and nutrient flows, common to all ecosystems when applied to the design of the human environment, provide the only ethical and pragmatic alternative to the future health of the emerging regional landscape. . . . where the role of technology is integrated with people, urbanism, and nature in ways that are biologically and socially self-sustaining and mutually supportive of life systems. These are the goals for shaping a new landscape based on fundamental environmental values (p. 194). . . . It is a perspective that is rooted in ecological and cultural diversity. If we look for it, the inherent potential for diversity shines like a beacon through the placeless dreariness of much contemporary urbanization. The making of memorable places involves principles of evolving natural process and change over time. It involves economy of means where often the less one does to make purposeful change the better. It involves variety and choice that evolve naturally through countless interactions between people and nature, providing a secure basis for ecological and social health (p. 210).

Appendix

RFP Rating Sheet For:

Item	Possible Points	Score
Planning		
Expertise in planning studies of area-wide water systems and traditional and non-traditional wastewater management systems.	30	
Expertise in environment assessments of water and wastewater management systems.	30	
Ability to conduct informative public meetings and obtain public involvement.	20	
Ability to prepare accurate cost estimates and revenue projections for water and wastewater projects.	20	
Familiarity with federal, state, and private funding sources.	10	
Subtotal Planning Services	110	
Design		
Expertise in design of water systems and traditional and non-traditional wastewater management systems.	30	
Expertise in development of standard plans and specifications for phased projects.	20	
History of accuracy of engineer's estimated construction cost with actual bid results.	10	
History of change orders related to design or specification deficiencies in construction projects.	10	
Expertise in aerial mapping and the subsequent preparation of construction plans.	10	
Subtotal Design Services	80	
Construction Management		
Expertise in the management of water and wastewater construction projects.	30	
Soundness of approach to construction management.	30	
History of final construction cost versus bid costs and time of completion versus original schedule.	10	
History of claims and claims resolution for construction projects.	10	
Subtotal Construction Management	80	
Total Score	270	

References

Adams, Lowell W. 1994. *Urban Wildlife Habitats—A Landscape Perspective.* Minneapolis: University of Minnesota Press.

Adams, Lowell W., and Louise E. Dove. 1989. *Wildlife Reserves and Corridors in the Urban Environment.* Columbia, Maryland: National Institute for Urban Wildlife.

Adams, Lowell W., and L. E. Dove. 1984a. *Urban Wetlands for Stormwater Control and Wildlife Enhancement.* Columbia, Maryland: National Institute for Urban Wildlife.

Adams, Lowell W., L. E. Dove, and D. L. Leedy. 1984b. Public attitudes toward urban wetlands for stormwater control and wildlife enhancement. *Wildlife Society Bulletin* 12: 299–303.

Adamus, P. R., and L. T. Stockwell. 1983. *A Method for Wetland Functional Assessment, Vols. 1 and 2.* Washington, D.C. Federal Highway Administration. FHWA-IP-82-23, 24.

Adamus, P. R., E. J. Clairain, Jr., R. D. Smith, and R. E. Young. 1987. *Wetland Evaluation Technique (WET), Volume II: Methodology.* Vicksburg, Mississippi. Department of the Army, Waterways Experiment Station.

Adamus, P. R., and K. Brandt. 1990. *Impacts on Quality of Inland Wetlands of the United States: A Survey of Indicators, Techniques and Applications of Community-Level Biomonitoring Data.* U.S. EPA Environmental Research Laboratory. EPA/600/3-90/073.

Alexander, Christopher. 1971. *A Pattern Language.* New York: Oxford University Press.

Allison, James. 1991. *Water in the Garden.* Boston: Little, Brown.

Archer-Wills, Anthony. 1993. *The Water Gardener.* London: Barron's Educational Series.

Bahr, T. G., R. C. Ball, and H. A. Tanner. 1974. The Michigan State University water quality management program, in *Wastewater Use in the Production of Food and Fiber—Proceedings.* U.S. EPA-660/2-74-041.

Best, E. P. H., S. L. Sprecher, H. L. Fredrickson, M. E. Zappi, and S. L. Larson. 1997. Screening submersed plant species for phytoremediation of explosives-contaminated groundwater from the Milan Army Ammunition Plant. Technical Report. U.S. Army Engineer Waterways Experiment Station, Vicksburg, MS.

Brodie, G. A., Donald A. Hammer, and David A. Tomljanovich. 1989. Treatment of acid drainage with a constructed wetlands at the TVA 950 coal mine, in *Constructed Wetlands for Wastewater Treatment.* Chelsea, MI: Lewis Publishers.

Bureau of Mines. 1988. *Mine Drainage and Surface Mine Reclamation.* Pittsburg, PA.

Campbell, Craig. 1986. *Aquatic Systems in Wastewater Treatment: Criteria for the Planning of Integrated Facilities.* Thesis, Department of Landscape Architecture, University of Washington, Seattle.

Carde, Margaret. 1990. Flowforms—Rx for Dying Water? in *Design Spirit,* Vol. II, No. 3, 1990.

Caduto, Michael J. 1985. *Pond and Brook.* Englewood Cliffs, NJ: Prentice-Hall.

Clemens, Herschel (trans). 1913. Sextius Julius Frontinus, water commissioner of Rome, AD 97, in *Frontinus and the Water Supply of the City of Rome.* New York: Longmans, Green.

Dacey, J. W. H. 1988. Internal winds in waterlilies: An adaptation for life in anaerobic sediments. *Science,* 210:1017–1019.

Doczi, Gyorgy. 1981. *The Power of Limits.* Boston, MA: Shambhala Publications.

Dramstad, W. E., J. D. Olson, and R. T. Forman. 1996. *Landscape Ecology Principles in Landscape Architecture and Land-Use Planning.* Washington, D.C: Island Press.

Durant, Will, and Ariel Durant. 1954. *The Story of Civilization; Our Oriental Heritage.* New York: Simon and Schuster, New York, pp. 133.

Durrell, Gerald. 1988. *A Practical Guide for the Amateur Naturalist.* New York: Alfred A. Knopf.

Ferguson, Bruce K. 1994. *Stormwater Infiltration.* Chelsea, MI.: Lewis Publishers.

Filion, F. L. 1983. The importance of wildlife to Canadians: Highlights of the 1981 national survey. Cat. No. CW66-62/1983E. Canadian Wildlife Service, Ottawa, Ontario.

Gearhart, R. A., and M Higley. 1993. Constructed open surface wetlands: The water quality benefits and wildlife benefits—City of Arcata, California. In Moshiri, 1993.

Geotechnical Fabrics Association. 1996. *Geotechnical Fabrics Report 1996 Specifiers Guide.*

Gersberg, R. M., B. V. Elkins, S. R. Lyon, and C. R. Goldman. 1986. *Role of Aquatic Plants in Wastewater Treatment by Artificial Wetlands. Water Resources,* Vol. 20, No. 3:363–368.

Giesecke, F. E. 1938. *Proceedings of the Twentieth Texas Water Works Short School,* College Station, Texas, Feb. 1938.

Godfrey, P. J., E. R. Kaynor, S. Pelczarski, and J. Benforado, eds. 1985. *Ecological Considerations in Wetlands Treatment of Municipal Wastewaters.* New York: Van Nostrand Reinhold.

Grant, William. 1995. Spontaneous abortions possibly related to ingestion of nitrate-contaminated well water—LaGrange County, Indiana, 1991–1994. *Morbidity Mortality Weekly Report,* Vol. 45, No. 26. Centers for Disease Control, Atlanta, GA.

Grillo, Paul Jacques. 1975. *Form, Function, and Design.* New York: Dover.

Gross, M. A. 1987. Assessment of the Effects of Household Chemicals Upon Individual Septic Tank Performances. Research Project G-1212-07. Fayetteville, AR. Arkansas Water Resources Research Center.

Hammer, Donald A. 1992. *Creating Freshwater Wetlands.* Chelsea, MI: Lewis Publishers.

Hammer, Donald A., ed. 1989. *Constructed Wetlands for Wastewater Treatment.* Chelsea, MI: Lewis Publishers.

Heriteau, Jacqueline, and Charles B. Thomas. 1994. *Water Gardens.* New York: Houghton Mifflin Company.

Herskowitz, J. 1986. Listowel Artificial Marsh Project Report. Ontario Ministry of the Environment, Water Resources Branch, Toronto.

Hickok, E. A., M. C. Hannaman, and N. C. Wenck. 1977. *Urban Runoff Treatment Methods. Volume 1—Nonstructural Wetland Treatment.* Washington, D.C. EPA-600/2-77-217.

Hines, Michael, and Sherwood C. Reed. 1992. Personal communication.

Horner, Richard R., Brian W. Mar, Dimitris E. Spyridakis, and Tzn-Siang Wang. 1982. *Transport, Deposition and Control of Heavy Metals in Highway Runoff.* Seattle: University of Washington.

Hough, Michael. 1984. *City Form and Natural Processes.* New York: Van Nostrand Reinhold.

———. 1990. *Out of Place: Restoring Identity to the Regional Landscape.* New Haven and London: Yale University Press.

Hull, Lynne. 1993. Grizedale Forest Sculpture Park, in *Sculpture Magazine Maquette,* May 1993.

Jackson, J. B. 1980. *The Necessity for Ruins.* Amherst, MA: The University of Massachusetts Press.

Jantrania, Anish. 1998. Personal communication.

Johanson, Patricia. 1992. *Art and Survival.* North Vancouver, B. C., Canada: Gallerie Publications.

Kadlec, Robert H., and Robert L. Knight. 1996. *Treatment Wetlands.* Boca Raton, FL: CRC Press.

Kappraff, Jay. 1991. Connections: *The Geometric Bridge Between Art and Science.* New York: McGraw-Hill.

Kent, Donald M. 1994. *Applied Wetlands Science and Technology.* Boca Raton, FL: CRC Press.

Livingston, Eric, and Ellen McCarron. 1991. *Stormwater Management—A Guide for Floridians.* Tallahassee, FL: Florida Department of Environmental Regulation.

Lyle, John Tillman. 1994. *Regenerative Design for Sustainable Development.* New York: John Wiley & Sons.

———. 1985. *Design for Human Ecosystems.* New York: Van Nostrand Reinhold.

Matilsky, Barbara C. 1992. *Fragile Ecologies: Artist's Interpretations and Solutions.* New York: Rizzoli.

McAllister, L. S. 1992. *Habitat Quality Assessment of Two Wetland Treatment Systems in Mississippi—A Pilot Study.* U.S. EPA. Environmental Research Laboratory, Corvallis OR. EPA/600/R-92/229.

McAllister, L. S. 1993a. *Habitat Quality Assessment of Two Wetland Treatment Systems in the Arid West—A Pilot Study.* U.S. EPA. Environmental Research Laboratory, Corvallis OR. EPA/600/R-93/117.

———. 1993b. *Habitat Quality Assessment of Two Wetland Treatment Systems in Florida—A Pilot Study.* U.S. EPA. Environmental Research Laboratory, Corvallis OR. EPA/600/R-93/222.

McCann, Kevin, and Lee Olson. 1994. *Final Report on Greenwood Urban Wetland Treatment Effectiveness.* Orlando, FL: City of Orlando Stormwater Utility Bureau.

McHarg, Ian. 1969. *Design with Nature.* New York. Natural History Press.

Mitsch, William J., and James G. Gosselink. 1993. *Wetlands.* New York: Van Nostrand Reinhold.

Mollison, William. 1988. *Permaculture: A Designer's Manual.* Tyalgum, Australia. Tagari Publications.

Moshiri, Gerald A., ed. 1993. *Constructed Wetlands for Water Quality Improvement.* Chelsea, MI.: Lewis Publishers.

Mumford, Lewis. 1952. *Art and Technics.* New York: Columbia University Press.

Odum, H. T. *Systems Ecology—An Introduction.* New York: John Wiley & Sons.

Olsen, Donald J. 1986. *The City as a Work of Art.* New Haven: Yale University Press.

Payne, Neil F. 1992. *Techniques for Wildlife Habitat Management of Wetlands.* New York: McGraw-Hill.

Phoenix Arts Commission. 1992. *Public Art Works: The Arizona Models.* Phoenix, Arizona: Phoenix Arts Commission.

Pineywoods Resource Conservation & Development, Inc. 1995. *The Feasibility of Constructed Wetlands in the Trinity River Watershed.* Nagogdoches, TX: Pineywoods Resource Conservation and Development.

Read, Herbert. 1961. *The Grass Roots of Art.* Cleveland, OH: World Publishing.

Reed, Sherwood C. 1993. *Subsurface Flow Constructed Wetlands for Wastewater Treatment—A Technology Assessment.* U.S. EPA, EPA832-R-93-008.

Reed, Sherwood E., et al. 1993. *Natural Systems for Wastewater Treatment.* New York: McGraw-Hill.

Reed, S. C., R. W. Crites, and E. J. Middlebrooks. 1995. *Natural Systems for Waste Management and Treatment,* 2nd ed. New York: McGraw-Hill.

Reid, George K. 1987. *Pond Life.* New York: Golden Press.

Riemer, Donald N. 1984. *Introduction to Freshwater Vegetation.* Westport, CT: Avi Publishing.

Schicker, L. 1987. Design criteria for children and wildlife in residential development, in L. W. Adams and D. L. Leedy, eds. *Integrating Man and Nature in the Metropolitan Environment.* Natl. Inst. for Columbia, MD: Urban Wildlife, pp. 95–105.

Schueler, Thomas R. 1992. *Design of Stormwater Wetland Systems.* Washington, D.C.: Dept. of Environmental Programs, Metropolitan Washington Council of Governments.

———. 1995. *Site Planning for Urban Stream Protection.* Silver Spring, MD: The Center for Watershed Protection and Metropolitan Washington Council of Governments.

Schwenk, Theodor. 1965. *Sensitive Chaos.* London: Rudolf Steiner Press.

Scow, Kate M. 1994. The Efficacy and Environmental Impact of Biological Additives to Septic Tanks. Report, University of California, Davis.

Seidel, Kathe, Helga Happel, and Georg Graue. 1976. *Contributions to Revitalisation of Waters,* Limnologische Arbeitsgruppe in der Max Planck Gesellschaft, Krefeld-Hulserberg. Germany.

Sloey, W. E., F. L. Spangler, and C. W. Fetter, Jr. 1978. Freshwater Wetlands for Nutrient Assimilation, in: R. E. Good, D. F. Whigham, and R. L. Simpson, eds. *Freshwater Wetlands.* New York: Academic Press.

Snow, John. 1857. *Snow on Cholera, being a Reprint of Two Papers by John Snow, M.D.,* New York, 1936; from a footnote in *Plagues and Peoples,* by William H. McNeill. 1974. New York: Anchor Press/Doubleday.

Spangler, Frederic L., William E. Sloey, and C. W. Fetter, Jr. 1976. *Wastewater Treatment by Natural and Artificial Marshes.* U.S. EPA-600/2-76-207, Ada, OK.

Steiner, Frederick. 1991. *The Living Landscape: An Ecological Approach to Landscape Planning.* New York: McGraw-Hill.

Steiner, Gerald R., and James T. Watson. 1993. *General Design, Construction, and Operation Guidelines: Constructed Wetlands Wastewater Treatment Systems for Small Users Including Individual Residences.* Chattanooga, Tennessee. Water Management Resources Group, Tennesee Valley Authority.

Stevens, Peter S. 1974. *Patterns in Nature.* Boston: Little, Brown.

Stolp, Heinz. 1988 *Microbial Ecology.* Cambridge, UK: Cambridge University Press.

Tchbanoglous, G., and F. L. Burton. 1991. *Wastewater Engineering. Treatment, Disposal, Reuse,* 3rd ed. New York: McGraw-Hill.

Tchobanoglous, G., and G. Culp. 1980. Aquaculture Systems for Wastewater Treatment: An Engineering Assessment. U.S. EPA, Office of Water Program Operations, Washington, D.C. EPA/430/9-80-007, NTIS No. PB 81-156689, pp. 13–42.

Thayer, R. L., and T. Westbrook. 1990. Open channel design systems for residential communities, in Proceedings of 1980 CELA conference, S. K. Williams and R. R. Grist, eds. Washington, D.C.: Landscape Architecture Foundation.

Thayer, Robert L., Jr. 1994. *Gray World, Green Heart.* New York: John Wiley & Sons.

The Guild. 1995. *The Guild Sourcebook of Artists Architect's Edition 10.* Madison, WI: Kraus Sikes.

Thompson, D'Arcy. 1961. *On Growth and Form.* London: Cambridge University Press.

Tourbier, J. Toby, and R. W. Pierson, Jr. 1976. *Biological Control of Water Pollution.* Philadelphia: University of Pennsylvania Press.

Tourbier, J. Toby, and R. Westmacott. 1981. *Water Resources Protection Technology: A Handbook of Measures to Protect Water Resources in Land Development.* Washington, D.C.: ULI–Urban Land Institute.

———. 1992. *Lakes and Ponds,* 2nd edition. Washington, D.C.: ULI–the Urban Land Institute.

———. 1980. *Manual On-site Wastewater Treatment.* U.S. EPA Office of Water, Cincinnati, OH. EPA625/1-80-012.

———. 1981. *Land Treatment of Municipal Wastewater.* U.S. EPA Office of Water, Cincinnati, OH. EPA625/1-81-013.

———. 1981. *Process Design Manual: Land Treatment of Municipal Wastewater.* Center for Environmental Research, Cincinnati, OH. EPA625/1-81-013.

U.S. EPA. 1988. *Design Manual: Constructed Wetlands and Aquatic Systems for Municipal Wastewater Treatment.* Cincinnati, OH: U.S. EPA. EPA/625/1-88/022.

———. 1992. *Wastewater Treatment/Disposal for Small Communities.* U.S. EPA Office of Research and Development, Cincinnati, OH. EPA625/R-92/005.

———. 1993. *Constructed Wetlands for Wastewater Treatment and Wildlife Habitat: 17 Case Studies.* U.S. EPA Office of Water, Cincinnati, OH. EPA832-R-93-005.

———. 1993a. *Guidance for Design and Construction of a Subsurface Flow Constructed Wetland.* Dallas, TX: U.S. EPA Region 6, Water Management Division.

———. 1993b. *Constructed Wetlands for Wastewater Treatment and Wildlife Habitat.* Washington, D.C.

———. 1993c. *Subsurface Flow Constructed Wetlands for Wastewater Treatment: A Technology Assessment.* U.S. EPA Office of Water. EPA 832-R-93-008. Municipal Technology Branch, U.S. EPA. EPA832-R-93-005.

———. 1993d. *Ground Water Currents.* EPA Series 542.

Wagenknecht-Harte, Kay. 1989. *Site + Sculpture.* New York: Van Nostrand Reinhold.

Warshall, Peter. 1979. *Septic Tank Practices.* Garden City, NY: Anchor Press/Doubleday.

Washington State Department of Ecology. 1992. *Stormwater Management Manual for the Puget Sound Basin.* Department of Ecology #91-75.

Weller, Milton W. 1994. *Freshwater Marshes: Ecology and Wildlife Management.* Minneapolis: University of Minnesota Press.

Wile, I., G. Miller, and S. Black. Design and use of artifical wetlands, in Godfrey, et al., 1985.

Wilson, Edmund O. 1998. *Consilience: the Unity of Knowledge.* New York: Alfred A. Knopf.

Wolverton, B. C. 1996. *How to Grow Fresh Air.* New York: Penguin Books.

———. 1989. *Aquatic Plant/microbial Filters for Treating Septic Tank Effluent,* in Hammer, 1989.

Wright, Frank Lloyd. 1910. *Ausgeführte Bauten und Entwürfe.* Berlin.

Reed, Sherwood C. 1990. *Natural Systems for Wastewater Treatment Manual of Practice FD-16.* Alexandria, VA: Water Pollution Control Federation.

Reed, Sherwood C., E. Joe Middlebrooks, and Ronald W. Crites. 1993. *Natural Systems for Waste Mangement & Treatment, 2nd edition.* New York: McGraw-Hill.

Crites, Ronald W., et al. 1988. *Constructed Wetlands and Aquatic Plant Systems for Municipal Wastewater Treatment.* Cincinnati, OH: U.S. EPA.

NOAA *Climatic Atlas of the U.S. 1988.* Washington, D.C.

Index